Cities and the Global Politics of the Environment

Series Editors
Michele Acuto
Melbourne School of Design
London, UK

Elizabeth Rapoport
Urban Land Institute
London, UK

Joana Setzer
The Grantham Research Institute on Climate
London School of Economics
London, UK

More than half of humanity lives in cities, and by 2050 this might extend to three quarters of the world's population. Cities now have an undeniable impact on world affairs: they constitute the hinges of the global economy, global information flows, and worldwide mobility of goods and people. Yet they also represent a formidable challenge for the 21st Century. Cities are core drivers not only of this momentous urbanisation, but also have a key impact on the environment, human security and the economy. Building on the Palgrave Pivot initiative, this series aims at capturing these pivotal implications with a particular attention to the impact of cities on global environmental politics, and with a distinctive cross-disciplinary appeal that seeks to bridge urban studies, international relations, and global governance. In particular, the series explores three themes: 1) What is the impact of cities on the global politics of the environment? 2) To what extent can there be talk of an emerging 'global urban' as a set of shared characteristics that link up cities worldwide? 3) How do new modes of thinking through the global environmental influence of cities help us to open up traditional frames for urban and international research?

More information about this series at
http://www.palgrave.com/gp/series/14897

Yonn Dierwechter

The Urbanization of Green Internationalism

palgrave
macmillan

Yonn Dierwechter
University of Washington Tacoma
Tacoma, WA, USA

Cities and the Global Politics of the Environment
ISBN 978-3-030-01014-0 ISBN 978-3-030-01015-7 (eBook)
https://doi.org/10.1007/978-3-030-01015-7

Library of Congress Control Number: 2018955460

This Palgrave Macmillan imprint is published by the registered company Springer Nature Switzerland AG
The registered company address is: Gewerbestrasse 11, 6330 Cham, Switzerland

Series Editors' Preface

A few years out of the world making grand promises about global sustainability, in agendas like the Paris Agreement on Climate Change or the Sustainable Development Goals, many are now on to seeking some proof we are in fact moving in the right direction. Key in several of these agendas, and in international politics more in general, has been the rise of cities and urban issues as important crossways to the achievement of a more sustainable future. Yet rarely we stop to appreciate what the "urban" really means in all of this.

Yonn Dierwechter's *The Urbanization of Green Internationalism* enters this picture at an apt moment: It argues that it is no longer possible to analyze global environmental politics without a more careful consideration of its various urbanization. In fact, it pluralizes this latter term and calls for an appreciation of the variety of urbanizations at play here in "green internationalism." As he points out in the book, the dynamics by which various institutions are reforming, reassembling, and adapting "up, down, and sideways" are an important entry point to understand the increasingly common references to the "urban" in world politics. In particular, the thrust behind the book is an importantly relational one: As Dierwechter advocates, we need to think about the new relationships between cities, states, and global environmental politics and how the various geographies of these relationships take shape in different cultural and historical settings, but also in their role as creators of new geographies. As he argues, while international affairs "have environmentalized, global environmentalism in turn has strongly urbanized in recent decades." Understanding what this urbanization means is a key issue for anyone working on sustainability *and* cities, and one that indeed could "in turn make us greener."

The Urbanization of Green Internationalism continues the series of investigations of this Pivot Series by calling for more systematic, historicized, and indeed geopolitical appreciation of these dynamics. It offers one more step in what we would argue is the right direction: treating cities and the various urbanizations underpinning the present "urban age" as essential determinants of the various elements of global governance.

London, UK

Michele Acuto
Joana Setzer
Elizabeth Rapoport

Acknowledgements

This book emerged originally from preparing lectures and facilitating discussions in a course I have taught for several years called "Green Internationalism and the City" offered in the Urban Studies Program at the University of Washington, Tacoma. I thank my colleagues for the opportunity to integrate this course in our curriculum on urban sustainability.

While teaching the course, I have invariably tried to expand urban studies beyond "local" policies and political dynamics to engage more broadly—if not always easily—with literature in international political economy, global environmental governance, and intellectual history. So additionally I want to thank my city-focused students for their admirable patience as I worked through—and indeed *still* work through—the ideas, arguments, and themes that follow here.

The book also shows, I hope, many years of teaching courses in political geography, planning theory, development studies, urban history, and metropolitan public policy. Although largely a work of geography, then, the discussion necessarily draws at different times on disciplines like political science, history, and planning. The case studies I deploy to illustrate key points furthermore exhibit the influence of a pre-academic career, international conferences, site visits, field work, consulting, and collaborations stretching back to the early 1990s.

Finally, I want to dedicate this little green book to my parents, Ron and Jewell Dierwechter, whose fifty-plus years of humanitarian internationalism opened up for their children (and now grandchildren) a world of people, places, and possibilities on multiple continents.

Contents

List of Figures

CHAPTER 1

Introduction: Cities, States, Global Environmental Politics

Abstract The recent rise of cities in global environmental politics has stimulated remarkable debates about sustainable urban development and the geopolitics of a changing world order no longer defined by tightly bordered national regimes. This book explores this major theme by drawing on approaches that document the diverse histories and emergent geographies of "internationalism." It is no longer possible, the book argues, to analyze the global politics of the environment without considering its various urbanization(s), wherein multiple actors are reforming, reassembling, and adapting to nascent threats posed by global ecological decay. The ongoing imposition and abrasion of different world orders—Westphalian and post-Westphalian—further suggests we need a wider frame to capture what the critical theorist of internationalism, Josep Antentas (Antipode, 47: 1101–1120, 2015), drawing on Daniel Bensaïd, calls the "sliding scale of spaces." The book will therefore appeal to students, scholars, and practitioners interested in global sustainability, urban development, planning, politics, and international affairs. Case studies and grounded examples of green internationalism in urban action presented later in the discussion ultimately explore how select city-regions are trying to negotiate and actually work through this postulated dilemma.

Keywords Global environmental politics · Urban geopolitics
Stateness · (Post)Westphalia

Y. Dierwechter, *The Urbanization of Green Internationalism*,
Cities and the Global Politics of the Environment,
https://doi.org/10.1007/978-3-030-01015-7_1

We have to deal simultaneously with the abrasion and the imposition of two orders: the modern (national) and the postmodern (cosmopolitan) reality.
—Pavel Frankowski (2010, p. 98)

Overview

The recent rise of cities in global environmental politics—the main theme of this Palgrave Pivot book series—has stimulated fresh, even remarkable, debates about the particular meanings and practices of sustainable urban development. It has challenged scholars, for instance, to think increasingly about the urban geopolitics of a more complex world order that is, by all accounts, no longer defined only (or mainly) by tightly bordered regimes of national regulation and social investment (Moisio, 2018; Yacobi, 2009). The rise of cities in global environmental politics has furthermore encouraged everyone—scholars certainly, but also politicians, public officials, citizens, and activists—to reconsider, reimagine, and reappraise the potential roles and actual contributions of the sovereign Westphalian state system that has for so long shaped our collective view of the world political map. New policy and political relationships between cities, states, and global environmental concerns like climate change have created nascent spatialities of sustainability no less than novel spatialities of cities, states, and global politics.

The rise of cities, though, *is* a debate. Some feel that pragmatic cities provide the best hope—the only plausible fix—for a global governance project that has stagnated through the dysfunction of nation-states (Barber, 2013, 2017). Exploring urbanism late in his career, Benjamin Barber (2013), a political theorist, optimistically amplified this renewed enthusiasm for the rise (or return) of cities in world development (e.g., Engelke, 2013). He offered a devastating critique of the nation-state, which he dismissed as increasingly incompetent, democratically distant, and emotionally attenuated from citizens. But the solution to the pathologies of the nation-state, he ultimately argued, is found in one of our oldest, most familiar inventions: the vibrant propinquity of the polis, or what Ed Soja (2000) in several books and multiple public lectures thought of as an urban "synekism" that has long forged geo-historic change and innovation (cf. Taylor, 2013). These were—and are—voices of urban(ized) hope. Global progress—not just *in* cities but *through* cities; through politically creative urban efforts to transform nature's bounty into socially just economies—endogenously within specific cities, of course, but especially across cities as they circulate ideas and lobby exogenous institutions for solutions to daunting problems. Barber's thesis, in particular, and certainly his overriding normative concerns, thus border on a kind of "city-statism," a claim/dream predicated on the idea/l that cities are, after all, older than states and altogether more likely in practice to articulate vibrant democracy with green efficacy.

Others wonder, however, just how cities (or city-regions) can unleash their presumed policy and project creativity in global affairs—in so far as they can or do—without the political support of a reformed, yet still institutionally relevant, kind of national state apparatus (Curtis, 2016). Arguing for "a new internationality," Herrschel and Newman (2017) are nonetheless careful to note, for example, that while cities are "international actors," they have *joined* rather than replaced states in forging a novel kind of global geopolitics and solution-seeking policy world. Cities and regions are going "beyond" the nation-state, in their estimation, but the state has hardly vanished. Indeed, Herrschel and Dierwechter (2018) have suggested more recently that efforts by key city-regions to balance urban competitiveness with global sustainability through new forms of "smart" regionalization are actually better understood as a "dual transition." One transition is "internal" to the politics of city-regions; the other, "external" to the city-region. Simon Curtis (2016, p. 456) further observes, quite compellingly in my view, that the rise of the city in world (environmental) politics "should not be seen as a symptom of the exhaustion of the state but rather as an adaptation of the state as it tries to cope with a changing environment."

The thesis of the state's "adaptation" is also central to the work of Andy Jonas and Sami Moisio (2016). They chart recent forms of city-regionalism as part and parcel of a new type of geopolitics rather than, say, an updated city-statism of disembodied globalized city-regions independently building a twenty-first-century version of the Delian, Lombard, or Hanseatic Leagues of different (urban) geopolitical pasts.[1]

[1]The Delian League, led by Athens, united about 300 Greek city-states in the fifth century BCE against the imperial threat of Persian rule. Similarly, the Lombard League was a military alliance of Northern Italian city-states—or merchant republics—in the Late Middle Ages and early Renaissance that protected members against the Italian kingdom and Holy Roman Empire. Finally, from the fourteenth to seventeenth centuries, the Hanseatic League of cities fielded its own militaries, constructed an effective exchange mechanism and tariff system, and sometimes met as a Parliament, known as the Hansetage, albeit infrequently. Each are examples of alternative ways of organizing supra-local geopolitical space-economies, which arguably today are also in considerable flux.

In particular, they suggest, efforts to sustain city-regions—to green up metropolitan areas—are deliberate strategies in how today's increasingly "polycentric states" now secure and project political power across the spaces of global governance. This is changing international relations, without question, but also the geographies of the state (and of cities). Simon Curtis (2016, p. 456) again helps us to sharpen up this point. The "nation-state," he argues, "is being reassembled into new forms." As societies have urbanized, we might say, so too have their states, though unevenly and in ways we do not really quite understand. The unsustainable urban is changing world politics, in this view, even as unsustainable world politics are changing what cities are, what they do, how they act, and how we theorize their spaces. For Moisio (2018, p. 1) in particular, "…cities—and attendant urban politics—have not been passive outcomes of state territorial formation processes but instead have occupied a pivotal role in the dynamic geopolitical processes of the nation-state."

Several key questions emerge. Can we count on the hypothesized creativity of cities (on their own or in still ill-formed horizontal networks) to revive the existentially non-negotiable project of global sustainability? As much as we fear the state, as much as we worry about its militarization, "bluntness," and regressive neoliberalization—about its powers *over us*—can we really jettison the state's capacities to help social collectives puzzle through shared problems (Skocpol, 1985)? For some observers, including both eco-radical municipalists and market libertarians, we can—and we should (Bookchin, 2014). But if we still do *need* the state, should we not spend more time thinking critically about how reassembled, adapted, polycentric, urbanized states—"smarter," "eco-states," for instance—might edify and work productively with increasingly creative, internationalized cities and their urbanized hinterlands (Backstrand & Kronsell, 2015; Dierwechter, 2017; Eckersley, 2004)? Finally and most importantly for my purposes here, as we think about the new relationships between cities, states, and global environmental politics how do we think about the various *geographies* of these relationships in different cultural and historical settings? Indeed, *what* new geographies are these developments now creating?

Evidence mounts daily that the global community—however defined—needs and looks to cities more than ever before (Mohieldin & Ijjasz-Vasquez, 2018). Taking this as the main empirical warrant and point of departure, this book accordingly explores different forms of urbanization that increasingly accompany what I call here "green internationalism." The book's title—the urbanization of green internationalism—specifically suggests three core arguments as initial rather than comprehensive or final reflections on the kinds of questions just broached.

The first argument is that it is no longer possible to analyze the global politics of the environment without a careful consideration of its various *urbanizations*. Traditionally, of course, urbanization refers to a long-term demographic process whereby once predominately rural populations increasingly become urban populations, although what constitutes "urban" is defined differently in different societies (e.g., 10,000 people in Sweden; 5000 in India). In consequence, the "level" of urbanization refers to the percentage of the total population that resides in places that a particular society—Swedish, Indian, Peruvian, Australian, etc.—calls cities. Relatedly, the "rate" of urbanization so defined is expressed as an annual percentage of overall change. In this book, however, the term urbanization refers instead to the urban expressions, diverse confluences, emplacements, and/or materializations of intertwined sets of socio-political and policy relations stretched out across differently territorialized scales of power, regulation, and agency. What is *urbanized*, in other words, is not simply people; it is the *emplaced* sets of (political) processes, some of which are "planetary" in nature, others not (Brenner, 2014).

The second argument of the book is, therefore, that the emerging *urbanizations* of global environmental politics are empirical reflections of how various institutions—including place-based actors, cities, states, and the modern interstate system—are reforming, reassembling, and adapting up, down, and sideways in legitimately novel ways to the multiple threats to economic accumulation and social cohesion posed by global ecological decay. New forms of *urbanized* international politics—or urban geopolitics—include city-to-city networks, like C40, but also new modes of state-to-state concerns, expressed through international urban programs like Local Agenda 21 in the 1990s and early 2000s and the UN's Sustainable Development Goals initiative more recently. In addition, nearly everyone's growing concerns with "smartness" or building "smart cities" through digital technologies and potentially more disaggregated modes of territorial governance and service provision are only adding to the palimpsest of urban spatial formations and power-geometries. In this sense, I share but also seek to spatialize Pavel Frankowski's (2010, p. 98) general insight that "We have to deal simultaneously with the abrasion and the imposition of two orders: the modern (national) and the postmodern (cosmopolitan) reality."

To do so, in consequence, I also want to engage conceptually with interesting new work on the histories, geographies, and future politics of "internationalism" as one way to capture the abrasion and imposition of the modern and the postmodern; the national and the cosmopolitan; the state and the non-state orders now putatively (re)shaping the ongoing greening of world development in diverse cultural and political settings. I want to argue, as a third contribution, for a variegated green internationalism that signals theoretically not a *temporal* succession from a bounded, modern, Westphalian world order to an unbounded, post-Westphalian, networked order; instead I want to advance what the critical theorist of internationalism, Josep Antentas (2015), mobilizing in particular the work of Daniel Bensaïd, calls the "sliding scale of spaces." Here cities hold multiple times, many stories, and endless possibilities, if not always obvious probabilities. Antentas notes that for most critical scholars, the concepts and projects of internationalism refer to "the cross-border collective action of subaltern groups" (p. 1102). But it might also refer, more expansively and usually less radically, as Perry Anderson originally had it, to "any outlook, or practice, that tends to transcend the nation towards a wider community, *of which nations continue to form the main units*" (p. 1101, emphasis added). Most cities are neither subaltern groups nor or they nations, though they may well engage with both. As they do so, some cities are becoming more politically important than they have been in centuries.

The urbanization of green internationalism is an initial attempt to capture a few of the surely many new, contested, variegated and now heavily urbanized geographies of dialectical entanglement increasingly associated with the surely many new *outlooks* and political *practices* around environmental concerns that *tend* to transcend national borders. Such dialectical geographies of entanglement suggest an overall metaphysical commitment that foregrounds "the perpetual resolution of binary oppositions" (Gregory, 2000, p. 172). Specifically, as Harvey (1996) has proposed, the perpetual (attempted) resolutions of binaries—e.g., national: cosmopolitan; global: local; state: non-state; city: nature—produces "things" (like cities) that are constituted out of flows, processes, and relations: "things" appear to us a "permanences," but—as an ontological matter—are just "instanciations" of multiple socio-natural processes that stretch unevenly across spaces, times, and scales in complex ways.[2] Because of their multiplex nature, because they are constituted out of variously scaled flows and relations, cities are profoundly creative places, as Ben Barber rightly noted; but for *exactly* the same reasons they are also "cauldrons" of contradiction, as Robert Beauregard (2018) has recently reminded us. The histories and likely futures of urbanized environments in diverse cultural settings are not only about creativity and innovation, after all, but also about profound exploitation and deep suffering.

Within the context of these opening themes and arguments, this book explores the impacts that new forms of urbanization are now having on global environmental politics as well as the influence of global environmental politics on urban policy and development practices. Using examples later on from major city-regions in Europe, North America, Australia, Africa, and Asia, the overall analysis draws on a host of new debates and claims within intellectual and international history, urban geography, global political economy, spatial planning, political ecology, smart urbanism, and comparative urban studies. As I highlight a new area of applied and normative action—a new geopolitical space I again call here green internationalism—I highlight the stagnated efficacy and recent regression of state-based international activity around global environmental problems, on the one hand, and an often creative but still constrained municipality-based internationalism that has largely emerged in the 2000s, on the other.

For the purposes of this book, then, the term green internationalism is roughly interchangeable with what others call global ecological politics. However, *green internationalism* signals my effort to move away from Westphalian modes of analysis and thinking stricto sensu, linking as well into forms of trans-local and cross-border outlooks and practices within and between cities, scientists, trade unions, business groups, social reformers, and so on. The Westphalian system, of course, refers to the taken-for-granted world political system that was established very slowly after 1648 as part of the Peace of Westphalia in Europe, wherein three normative principles constitute the modern global system: (1) state sovereignty;

[2]While Harvey's own interpretation of dialects emphasizes the importance of class relations, ecologists also insist on flow and the hybrid nature of "things" (Barash, 2001). For their part too, post-structuralists highlight the dialectics of multiple forces, notably race, gender, sexuality, and ethnicity in the constitution or "thingification" of the material-ideational world.

(2) legal equality of states; and (3) non-intervention of one state in the affairs of another. A new set of processes, I suggest here, are indeed challenging the territorial and political anatomy of the Westphalian order, evident since the 1970s according to some researchers. This transformation is an uneven, inchoate, but increasingly legible shift from a *Westphalian* to a *post-Westphalian* world order in which international organizations (of various kinds) are becoming independent sites of authority and action, including urban policy networks. But in my judgment, this is more a *spatialized* process of abrasion and imposition between multiple orders—state and non-state; hegemonic and subaltern; old and new—than a *temporal* successful of one for the other. The urban geopolitics of the new green internationalism are not simply what cities *do with one another* in global space; they are also *what states do* (or don't do) with their cities and *what cities do* (or don't do) with their states (Jonas, 2013). The geopolitics of the global environment constitute one crucial arena in which to understand these urbanization dynamics.

In truth, they are tantalizingly more than that. The increasingly desperate global search for sustainability—defined throughout this book as the economic transformation of nature into ecologically resilient, democratically vibrant, and socially just societies—actually may be one of the principal reasons for this major development in world affairs in the first place. With a few obvious exceptions—like thermonuclear warfare, a super-volcano, or a large asteroid strike—the slow burn of global (un) sustainability since the industrialization and carbonization of capitalism first started in mid-eighteenth century Britain is easily the most significant threat to human life on earth as presently understood. As human life is now mostly and for some entirely urbanized, the sustainability of cities and the sustainability of the planet are increasingly the same problem.

Plan for the Book

Naturally, the analysis on offer here seeks to fit broadly within—and in fact is partly inspired by—this Palgrave Pivot book series edited by Acuto, Rapaport, and Setzer. As the co-editors note, the series generally tries to explore:

> the impact of cities on global environmental politics, and with a distinctive cross-disciplinary appeal that seeks to bridge urban studies, international relations, and global governance. The series explores three themes: 1)

What is the impact of cities on the global politics of the environment? 2) To what extent can there be talk of an emerging "global urban" as a set of shared characteristics that link up cities worldwide? 3) How do new modes of thinking through the global environmental influence of cities help us to open up traditional frames for urban and international research?

In reflecting upon these synoptic themes and developing the specific arguments and modes of empirical analysis of this book, the overall narrative that follows works "backwards" and roughly geo-historically. The overarching logic of the narrative, in other words, is first to engage and defend the concept of internationalism in Chapter 2. Chapter 3 then turns to the relatively recent greening (or environmentalization) of the various projects of internationalism. Next and probably most importantly from an empirical perspective, Chapter 4 discusses the even more recent urbanization(s) of this now greened-up internationalism. Finally, the book recapitulates the main arguments and addresses future research questions in Chapter 5.

Specifically, Chapter 2 engages literatures in various disciplines on the intellectual histories and always contested geographies of internationalism, tracing the story with Mark Mazower and others to the early nineteenth century. Mazower's (2012) work, along with research by Brenda Sluga and several others, is important because it shows the relatively early emergence and institutional *heterogeneity* of various forms of internationalism located *outside* the traditional politics and diplomacy of the inter-state system per se (Sluga & Clavin, 2017). The internationalism of nations—of governing the world through a state-created framework—forms an important development of the older Westphalian order just discussed; but Mazower and others draw attention to numerous outlooks and practices, including what Mazower calls various, contending "brotherhoods" around free-market ideals, labor solidarities, evangelism, anti-slavery campaigns, legal codification, and scientific knowledge—all of which are surprisingly relevant today. That is why I retain and deploy the concept of internationalism, despite its territorial baggage.

Geographers have also charted the heterogeneity of internationalism, but particularly the transgressions that challenge hegemonic forms of power and oppression (Barchiesi, 2001; Featherstone, 2012, 2015; Ferretti, 2018). Resisting the periodization of time, or what might be considered the bookending of "orders" separated too cleanly from one another (Orren & Skowronek, 1996), my discussion here instead suggests that both Westphalian and post-Westphalian (or *counter*-Westphalian) forms of internationalism have *long* co-shaped global politics—and thus are likely to do in the future as cities gain even more clout. This ultimately requires us to reframe global political narratives that tend to neglect non-state spatialities and especially subaltern forms of global or cross-border agency (Herod, 2004).

Chapter 3 explores moments in the rise of, and eventual stagnation associated with, the greening of internationalism. Major concerns include the growing recognition of the environment in world political discourses and deepening concerns with threats posed by irreversible ecological limits across various global commons. The chapter outlines select political efforts to address globally shared environmental problems over several decades of multilateral diplomacy (e.g., Broadhead, 2002). In addition, the discussion traces the recent stagnation and reversal of the traditional Westphalian order in addressing multiple green challenges to society and economy (Amen, Toly, McCarney, & Segbers, 2012; Barber, 2013).

Chapter 4, the longest and most important in the book, builds on the conclusions of Chapter 3. The discussion notes that while international affairs have environmentalized, global environmentalism in turn has strongly *urbanized* in recent decades. The chapter highlights tensions that major urban changes have placed on inherited structures of global policy space, such as the early importance of the Rio Earth Summit, in general, and Local Agenda 21, in particular, in elevating the presumed pragmatism of cities. The discussion presses the case for how urban space was steadily reconceptualized after the denouement of the Cold War as a global solution to ecological challenges. One major implication is that political ecologies have now delocalized and upscaled, a process that has caused its own tensions and political contradictions, made clearer by the regressive populism of 2016, but long-simmering with the impacts of neoliberal globalization (Dierwechter, 2008; Trapenberg Frick, 2013). Attention is paid in this chapter to signature initiatives like Local Agenda 21, Sustainable Development Goals (SDGs) , the rise of inter-municipal policy networks, and recent "smartness" discourses.

As a narrative strategy, Chapter 4 deploys the synoptic metaphor of "romance" to assess the rise of cities as they engage states and international organizations, and vice versa. The "romancing of cities," the narrative suggests, could lead to important new relationships or, alternatively, to a dreamy romanticization of unlikely futures. Examples seek to elaborate on emerging urban manifestations of green internationalism as distinct forms of global geopolitics, building on insights and claims by various scholars whose work informs—explicitly or implicitly—the synoptic themes posed by this Palgrave Pivot series (Acuto & Rayner, 2016; Bouteligier, 2012, 2013; Herrschel & Newman, 2017; Jonas, 2013; Lee, 2013; Parnell & Robinson, 2012; Taylor, 2004).

The chapter attempts to foreground the empirical emergence of three (still new) forms or expressions of globalized urbanization, namely: *international urbanization*, focusing especially on the urbanization of the United Nations; *transnational urbanization*, focusing on inter-city networks such as C40; and perhaps most recently, *smart city urbanization*, focusing on the generalized infatuation with new Internet and communication (ICT) technologies of digitalization at various territorial scales and modes of governance. Each urban expression critically synthesizes recent literatures from multiple societies and explicitly highlights an illustrative case study, including Cape Town, Los Angeles, and Melbourne, respectively.

Finally, Chapter 5 recapitulates the main ideas of the preceding chapters and assesses the limitations of the arguments from both a normative and analytical perspective. The book ends with a brief discussion of a possible research questions going forward, particularly as this agenda involves a comparative urban studies program increasingly sensitive to the shared geopolitics of environmentalism. Such a program should consider discontent with green policies that are linked discursively to elite actors putatively hostile to industry, nationalism, working-class lives, and "local" democracy (Trapenberg Frick, 2013), but which arguably now constitute a (once unexpected) part of what Brauch, Dalby, and Spring (2011) have called the new political geoecologies of the Anthropocene. Here again, and for the final time, the book reconsiders how the "abrading and imposition" of different world orders are co-shaping the contemporary interplay between cities, states, and global environmental politics.

References

Acuto, M., & Rayner, S. (2016). City networks: Breaking gridlocks or forging (new) lock-ins? *International Affairs, 92*(5), 1147–1166. https://doi.org/10.1111/1468-2346.12700.

Amen, M., Toly, N., McCarney, P., & Segbers, K. (Eds.). (2012). *Cities and global governance: New sites for international relations.* Farnham, UK: Ashgate.

Antentas, J. (2015). Sliding scale of spaces and dilemmas of internationalism. *Antipode, 47*(5), 1101–1120. https://doi.org/10.1111/anti.12169.

Backstrand, K., & Kronsell, A. (2015). *Rethinking the green state: Environmental governance towards climate and sustainability transitions*. London and New York: Routledge.

Barash, D. (2001, February 23). Buddhism and the subversive science. *Chronicle of Higher Education*. Retrieved from https://www.chronicle.com/article/Buddhismthe-Subversive/32948.

Barber, B. (2013). *If mayors ruled the world: Dysfunctional nations, rising cities*. New Haven: Yale University Press.

Barber, B. (2017). *Cool cities: Urban sovereignty and the fix for global warming*. New Haven, CN: Yale University Press.

Barchiesi, F. (2001). Transnational capital, urban globalisation and cross-border solidarity: The case of the South African municipal workers. In P. Waterman & J. Wills (Eds.), *Place, space and the new labour internationalisms*. Oxford, UK; Malden, MA: Blackwell.

Beauregard, R. (2018). *Cities in the urban age: A dissent*. Chicago: University of Chicago Press.

Bookchin, D. (2014). *Radical municipalism: The future we deserve*. Retrieved from https://www.filmsforaction.org/articles/radical-municipalism-the-future-we-deserve/.

Bouteligier, S. (2012). *Global cities and networks for global environmental governance*. Hoboken: Taylor and Francis.

Bouteligier, S. (2013). Inequality in new global governance arrangements: The north-south divide in transnational municipal networks. *Innovation—The European Journal of Social Science Research, 26*(3), 251–267. https://doi.org/10.1080/13511610.2013.771890.

Brauch, H., Dalby, S., & Spring, Ú. (2011). Political geoecology for the anthropocene. In H. G. Brauch, Ú. Oswald Spring, C. Mesjasz, J. Grin, P. Kameri-Mbote, B. Chourou, … J. Birkmann (Eds.), *Coping with global environmental change, disasters and security* (Vol. 5, pp. 1453–1485). Berlin, Heidelberg: Springer.

Brenner, N. (2014). *Implosions/explosions: Towards a study of planetary urbanization*. Berlin: Jovis.

Broadhead, L.-A. (2002). *International environmental politics: The limits to green diplomacy*. Boulder, CO: Lynne Rienner.

Curtis, S. (2016). Cities and global governance: State failure or a new global order? *Millennium, 44*(3), 455–477. https://doi.org/10.1177/0305829816637233.

Dierwechter, Y. (2008). *Urban growth management and its discontents: Promises, practices and geopolitics in US city-regions*. New York: Palgrave.

Dierwechter, Y. (2017). The smart state as utopian space for urban politics. In A. Jonas, B. Miller, K. Ward, & D. Wilson (Eds.), *The Routledge handbook on spaces of urban politics*. London and New York: Routledge.

Eckersley, R. (Ed.). (2004). *The green state*. Boston: MIT Press.

Engelke, P. (2013). *Foreign policy for an urban world: Global governance and the rise of cities*. Washington, DC: Atlantic Council.

Featherstone, D. (2012). *Solidarity: Hidden histories and geographies of internationalism*. London and New York: Zed Books.

Featherstone, D. (2015). Maritime labour and subaltern geographies of internationalism: Black internationalist seafarers' organising in the inter-war period. *Political Geography, 49*, 7–16. https://doi.org/10.1016/j.polgeo.2015.08.004.

Ferretti, F. (2018). Geographies of internationalism: Radical development and critical geopolitics from the Northeast of Brazil. *Political Geography, 63*, 10–19. https://doi.org/10.1016/j.polgeo.2017.11.004.

Frankowski, P. (2010). One world-many orders? In S. Krishna-Hensel (Ed.), *Order and disorder in the international system*. Farnham: Taylor and Francis.

Gregory, D. (2000). Dialectic(s). In R. Johnston, D. Gregory, G. Pratt, & M. Watts (Eds.), *The dictionary of human geography* (4th ed., pp. 172–173). Oxford: Blackwell.

Harvey, D. (1996). *Justice, nature and the geography of difference*. Oxford: Blackwell.

Herod, A. (2004). Geographies of labor internationalism. *Social Science History, 27*(4), 501–523. https://doi.org/10.1017/s0145553200012669.

Herrschel, T., & Dierwechter, Y. (2018). *Smart transitions in city-regionalism: The quest for competitiveness and sustainability*. London: Routledge.

Herrschel, T., & Newman, P. (2017). *Cities as international actors: Urban and regional governance beyond the nation state*. London and New York: Palgrave.

Jonas, A. (2013). City-regionalism as a contingent 'geopolitics of capitalism'. *Geopolitics, 18*(2), 284–298. https://doi.org/10.1080/14650045.2012.723290.

Jonas, A. & Moisio, S. (2016). City regionalism as geopolitical processes: A new framework for analysis. *Progress in Human Geography*. https://doi.org/10.1177/0309132516679897.

Lee, T. (2013). Global cities and transnational climate change networks. *Global Environmental Politics, 13*(1), 108–127. https://doi.org/10.1162/GLEP_a_00156.

Mazower, M. (2012). *Governing the world: The history of an idea, 1815 to the present*. New York: The Penguin Press.

Mohieldin, M., & Ijjasz-Vasquez, E. (2018). *Localizing the SDGs in Colombia, Indonesia and Kenya*. Retrieved from http://www.citiscope.org/commentary/2017/10/localizing-sdgs-colombia-indonesia-and-kenya.

Moisio, S. (2018). Urbanizing the nation-state? Notes on the geopolitical growth of cities and city-regions. *Urban Geography*, 1–4. https://doi.org/10.1080/02723638.2018.1454685.

Orren, K., & Skowronek, S. (1996). Institutions and intercurrence: Theory building in the fullness of time. *Nomos XXXVII, Political Order, 38*, 111–146.

Parnell, S., & Robinson, J. (2012). (Re)Theorizing cities from the global south: Looking beyond neoliberalism. *Urban Geography, 33*(4), 593–617.

Skocpol, T. (1985). Introduction. In P. Evans, R. Rueschemeyer, & T. Skocpol (Eds.), *Bringing the state back in* (pp. 3–38). Cambridge and New York: Cambridge University Press.

Sluga, G., & Clavin, P. (2017a). Rethinking the history of internationalism. In G. Sluga & P. Clavin (Eds.), *Internationalisms: A twentieth century history* (pp. 3–16). Cambridge: Cambridge University Press.

Soja, E. (2000). *Postmetropolis*. London and New York: Sage.

Taylor, P. J. (2004). *World city network: A global urban analysis*. London and New York: Routledge.

Taylor, P. J. (2013). *Extraordinary cities: Millenia of moral syndromes, world-systems and city/state relations*. Cheltenham: Edward Elgar.

Trapenberg Frick, K. (2013). The actions of discontent: Tea party and property rights activists pushing back against regional planning. *Journal of the American Planning Association, 79*(3), 190–200. https://doi.org/10.1080/01944363.2013.885312.

Yacobi, H. (2009). Towards urban geopolitics. *Geopolitics, 14*(3), 576–581. https://doi.org/10.1080/14650040802694091.

CHAPTER 2

Contending Internationalisms: Times, Spaces, Frames

Abstract This chapter engages with recent work on the histories and geographies of "internationalism," tracing the story back to the early nineteenth century. While the internationalism of nations—of governing the world through a state-created framework—forms an important development of the older Westphalian order mentioned in Chapter 1, this chapter also draws attention to other outlooks and practices, including what Mark Mazower (Governing the world: The history of an idea, 1815 to the present. The Penguin Press, New York, 2012) calls diverse "brotherhoods" around free market ideals, evangelical moralism, anti-slavery campaigns, legal codification, and scientific knowledge as well as to what geographers have similarly charted as the heterogeneity of subaltern movements that challenge hegemonic forms of power and oppression (Featherstone, in Solidarity: Hidden histories and geographies of internationalism. Zed Books, London, New York, 2012; Political Geography, 49: 7–16, 2015; Ferretti, in Political Geography, 63: 10–19, 2018). Resisting the periodization of time, the chapter suggests that both Westphalian and post-Westphalian (or perhaps *counter*-Westphalian) forms of internationalism have *long* co-shaped global politics—and are therefore likely to do so in the future as cities gain even more clout. This ultimately requires us to reframe global political narratives that erase non-state spatialities and forms of agency, whether past, present, or future (Herod in Social Science History 27: 501–523, 2004).

Y. Dierwechter, *The Urbanization of Green Internationalism*, Cities and the Global Politics of the Environment, https://doi.org/10.1007/978-3-030-01015-7_2

Keywords Internationalism · Scale · Space · (Neo)Liberal order
Rise of cities

> *In the twenty-first century, historical interest in internationalism – as anything but the call to the workers of the world to unite – is gradually becoming the norm in a relatively short space of time.*
>
> —Glenda Sluga and Patricia Clavin (2017a, p. 4)

Introduction

This chapter makes an initial case for how to think about the emerging relationships between cities, states, and global environmental politics through the organizing concept of internationalism. Internationalism per se is usually first associated with the movement politics of socialism, in general, and the actual and/or potential solidarities of labor across borders, in particular (Forman, 1998). But the outlooks and practices of internationalism have also referred to other forms of political agency and planetary-wide activities, including, as discussed briefly in Chapter 1, the internationalism of modern nation-states as well as the internationalisms of various "subaltern" groups (Featherstone, 2012). As Sluga and Clavin (op cit.) observe in the head quote above, historians of both intellectual and pragmatic developments—from the advocacy of free trade to indigenous rights campaigns—have recently rediscovered older forms of internationalism that, arguably, are still playing out today, albeit in altered and perhaps unexpected forms.

Geographers, though long characterized by "a puzzling silence" (Hodder, Legg, & Heffernan, 2015, p. 2), have grown increasingly interested in using the concept, multi-scalarity, and political projects of internationalism to (re)explore, inter alia, the spatialities, interests, and/or solidarities of subaltern races and ethnicities; market liberalism; labor exploitation; thirdworldism; peacemaking; suffragette rights; religious diasporas; legal codification; counter-globalization; and critical knowledge networks (Brooks, 2015; Chatterton, Featherstone, & Routledge, 2013; Featherstone, 2012; Hodder et al., 2015). Much of this geographical work is often historical in nature, creating synergies across two disciplines long important in comparative urban studies (Brooks, 2015; Ferretti, 2018; Herod, 2004). We benefit from, as

David Featherstone (2012, p. 46) observes, "a plural and generous account" of internationalism. The extent to which internationalism is now changing—and how—is a major theme in multiple fields, opening up fresh terrain, in my judgment, to reconsider the role of cities in the global political order.

Internationalism so expansively conceived requires nuanced treatments of time, scale, and space, particularly where questions of the state's presumed pasts and possible futures are concerned. As also discussed in Chapter 1, the state-based international system emerged slowly after the famous Peace of Westphalia in 1648. But the development of internationalism—considered by Mark Mazower (2012) as seeing the world as a single governable space—is much more recent, dating to the initially conservative and Eurocentric political order created at the Concert of Europe in the aftermath of the Napoleonic Wars in 1815. In this sense, internationalism as both *outlook* and *practice* is more historically complex, geographically striated, and politically open than the steady march of increasingly integrated and more expansive state-based diplomatic activity might otherwise suggest—as important as *that* particular reality has been to the emergence of the modern political world. Put differently, multiple internationalism*s* emerged during the nineteenth century, even as the state-based system—the Westphalian system, as I call it throughout this book—remained hegemonic across the long twentieth century, and even today to a large extent.

Because multiple histories and geographies of internationalism characterize the past, multiple histories and geographies of internationalism still shape the present—and likely the future, too (Featherstone, 2012). These include new urban forms of green internationalism based on the increasingly well-documented activities of cities actively searching for (and experimenting with) novel ways of "doing" global politics and policy learning within, through, and sometimes beyond the formal architectures of a territorialized interstate decision-making apparatus. What we perhaps now need, then, is a way to organize and make sense of these variegated stories—of the pasts, presents, and futures—using different kinds of "frames" that generate different epistemologies of internationalism. My main purpose in what follows below is to develop this new thinking by tracing the geo-historical heritages and possible futures of various, often contending forms of internationalism.

From Past to Future: Times, Spaces, Scales

It is not quite accurate to assert in most circumstances that people live in the past. However, the past always lives in people—in their cultural norms, institutional designs, ethics and mores, and even future aspirations (Sowell, 1998). The past is never really over. Moreover, the past—history, heritage—is not a single *thing*, but more a tangled collection of temporalities that unfold unevenly over spaces and across scales. We should not speak of time *passing*, but of *times* passing, and even more accurately, of variegated space-times playing out over multiple scales.

Many scholars in different fields have developed this difficult theme. Daniel Bensaïd's synoptic notion of the "sliding scale of spaces," especially as applied to work on critical geographies of internationalism by Josep Antentas (2015) and others, is a good example. Bensaïd insisted on a "plurality of temporalities whose assemblage is contradictory and non-harmonious" (p. 1103). Various domains of life—economic, linguistic, military, and ecological, etc.—are increasingly disjointed in space as their respective "times" are multiple and "syncopated" and each projected (or stretched) at different scales (Antentas, 2016). This makes coordination difficult. Henri Lefebvre (2013) made similar points in his attempt to provide a "rhythmanalysis" of global capitalism. Geographers of time, such as Nigel Thrift, Alan Pred, and Torsten Hagerstrand, have also insisted on distinguishing the ways in which different *kinds* of times are place-entangled, rather than seeing time per se simply as the chronological march of years from a "before" to an "after" (Pred, 1977). In a different theoretical vein, historical institutionalists like Steve Skowronek and Karen Orren (2004), working in the neo-Weberian field of American Political Development (APD), have similarly argued for the "intercurrence" of multiple political "orders" at any given moment in any given place. We do not typically live in one political order; multiple orders live in us (for more see Dierwechter, 2017).

Such rethinking of times, spaces, and scales—and especially of contending multiple orders—helps us to understand emerging relationships between cities, states, and global environmental politics and thus important new conversations that have also emerged in parts of comparative urban studies, international relations, and political-economic scholarship on global sustainability (cf. Amen, Toly, McCarney, & Segbers, 2012; Attwell, 2014; Schroeder & Bulkeley, 2009).

Urban studies, to start with, has broken fully through the shackles of an inveterate localism (Ruffin, 2013). In recent decades, numerous scholars have shown that cities have profoundly globalized just as a globalized world has rapidly urbanized. Global cities, Saskia Sassen (1991) first argued in the early 1990s, represent a novel kind of historic space-economy: strategic command-and-control posts—New York, Tokyo, and London—crucial to organizing (and ensuring) the distanciated, border-piercing, dynamic of worldwide capital accumulation. In the wake of Sassen's work, urbanists of various kinds soon moved from opening up global cities per se to chronicling and making sense of the hypothesized globalization of nearly all cities (Czarniawska-Joerges, 2002). Jenny Robinson (2006) pushed this urban-cartographic flattening out still further, insisting that because all cities are globalized—and endlessly creative—all cities are also ordinary. Most recently, heady debates now swirl around Neil Brenner and Christian Schmitt's thesis of "planetary urbanization" (Brenner, 2014; Merrifield, 2013).

We can no longer step outside of the urban, Brenner and Schmitt specifically and provocatively claim, because urbanization's effects and flows stretch beyond cities as core places. They stretch beyond the spatially extended and intricately connected archipelago of global city-regions that Allen Scott (2001) first detected as part of the new spatial grammar of political economies in the twenty-first century. Following Henri Lefebvre's influential work on the social production of space, planetary urbanization means, for Brenner and Schmitt if hardly for everyone (e.g., Walker, 2015), that we must "reimagine the urban," as Amin and Thrift (2002) once put it. It means that we must reimagine and urbanize global political space itself, a theme also taken up by Peter Taylor (2005) and John Agnew (2000) in their work on the embryonic political geographies attending global capitalism. Finally, as Noah Toly (2008, p. 342) reminds us in ways that I shall revisit in this book: "While a vast literature is focused on the question, 'What is globalization doing to cities?', the flipside of that question—'What are cities doing to globalization?' or 'What are cities globalizing?'—is slightly less often engaged." All of this work rethinks, albeit in different ways, the spatialities and scalar assumptions of "the urban." It further undermines key temporalities of development, as places are drawn together dialectically.

Meanwhile, at least some scholars of international affairs and the global geopolitical economy have broken through their own temporal, spatial, and scalar shackles. These shackles are characterized primarily by the principles and practices of the Westphalian world order discussed earlier. This order has built up gradually over the past three-and-a-half centuries after the Peace of Westphalia (1648), which ended the bloody and costly Wars of Religion in Europe. The Westphalian world order spread out in theory from Europe through worldwide colonialism and eventually assumed conceptual hegemony in the presumed conduct of modern international affairs after the Napoleonic Wars. According to Jayabrata Sarkar (2015), the Westphalian system has long championed, if by no means entirely occasioned or practically ensured, key normative aspirations for a singular global political order:

> First, it secularised international politics by divorcing it from any particular religious footing, anchoring it instead on the tenets of national interest and reasons of state. Second, it promoted sovereignty, the legal doctrine that no higher authority stands above the state, except that to which the state voluntarily assents. Third, it accepted a conception of international society based on the legal equality of states. All sovereign states possessed the same rights and duties. They had the right to manage matters within their boundaries without outside interference, as well as the duty to abstain from intervening in the domestic affairs of other states. Thus, the Westphalian treaty [in 1648] marks the birth of the nation-state, itself the primary subject of modern international law. (n.p.)

A corpus of work in international affairs, also responding to the multi-dimensional impacts of economic and cultural globalization on both state and society, raises additional questions about the extent and meaning of a putative post-Westphalian (if not necessarily post-state) turn in world politics (Krishna-Hensel, 2010; Newman, 2009). A key concern in this work is the hypothesized erosion of Westphalian state sovereignty and especially the enervated capacities of even the strongest states "to manage matters within their boundaries" (Sarkar, op cit.). A related concern is with the pace, pathways, and potentialities of various kinds of non-state actors, from activist groups to global corporations, who are also engaged in transnational activities and projects—from human rights struggles to trade policy (Noortmann, Reinisch, & Ryngaert, 2015).

What is important, however, is to bring these various strands of thinking together, to account for the recent urbanizations of green internationalism from above and from below. Again, Pavel Frankowski (2010, p. 98) writes that "We have to deal simultaneously with the abrasion and the imposition of two orders: the modern (national) and the postmodern (cosmopolitan) reality." We live in "one world," he suggests, that is governed by "two world orders," whose interrelationships are unevenly unfolding through time, across scale, and over space. The histories, geographies, and futures of internationalism, as I shall now show, therefore means attending to the sliding scale of spaces and the imposition of multiple orders—orders created by traditional states, of course, but also by other actors working together to build new politics outside the territorial architectures of formal institutional spaces. It means attending to the production of various internationalisms and to traditions of "intercurrence" that live on today—and thus live in us.

Geo-traditions of Internationalism

Internationalism in this book refers in its broadest terms to the ideas and sociopolitical practices associated with what the intellectual historian Mark Mazower (2012) sees as governing the world. More specifically, less simplistically, and once again following lines traced by Perry Anderson and others, internationalism refers to "any outlook, or practice, that tends to transcend the nation towards a wider community, *of which nations continue to form the main units*" (quoted in Antentas, 2015, p. 1101, emphasis added).

Anderson's definition is useful because it includes but goes beyond nations and their states as the only actors (or "units", as he puts it) in global affairs. In consequence, while nations "continue to form the core" of the outlooks and practices of internationalism, they are not today—nor have they been in the past—alone in in search of "a wider community." Efforts by nations and their states to govern the world formally between themselves, in other words, have contended (and still contend) with other types of "cross-border collective action"—each producing what the geographer David Featherstone calls "the hidden geographies of subaltern groups that cover hidden stories of internationalism" (ibid., p. 1102). As also just discussed in Chapter 1, these dynamics help to explain and give cartographic shape to the "abrasion" and "imposition" associated with "the modern (national) world order and a postmodern (cosmopolitan) reality" (Frankowski, 2010, p. 98). Paradoxically, however, internationalism is nearly always conceptualized, by its proponents at least, as the crucial route to "a more peaceful, progressive or prosperous future" (Hodder et al., 2015). Ironically, fascists have had their own versions of internationalism; internationalism thus cannot be taken on its own terms. Steffek (2015) shows, for instance, that proposed blueprints for international institutions have been associated with a variety of political ideologies, from fascism to liberalism in addition to socialism.

In Mazower's view, the overall history of cross-border collective action, or what he also calls the story of global cooperation, evolved over the nineteenth and twentieth centuries, initially as a set of multi-dimensional responses to efforts by European states in 1815 to reestablish conservative stability at the Congress of Vienna in the revolutionary and (for the elites at least) rather unnerving wake of the French Revolution and Napoleonic Wars. Renewed political ideas like democracy, human rights, and constitutionalism—which had occasionally appeared before in political history but subsequently vanished or weakened in early modernity—now threatened conservative concepts of monarchy shattered by war and cultural modernity.

The economic rise of the urban commercial class similarly challenged the divinely ordained rights of royal families to rule societies. While the post-Napoleonic Concert of Europe provided what Mazower further calls the first "model" of international governance, it nonetheless set in motion a series of counter-reactions and alternative imaginaries of internationalism which, "as a [program] of cooperation among nations and their peoples, was moving from the realm of marginal ideas into the mainstream" (p. 23). (The quite recent urbanization of this program's sustainability turn—and where this turn may now be heading—in the late twentieth and early twenty-first centuries is a principal analytical concern later in this book.)

Internationalism is relevant again. For the past thirty years, globalization—whether mapped as neoliberal or empowering—has arguably sidelined intellectual discourses of internationalism; these include but are not limited to radical class-based traditions of labor (Imlay, 2017). In one sense, this is hardly a problem. Globalization, after all, also effectively denationalizes and decenters the territorial hegemony of "the nation" in ways that open much needed space for other identities, societal projects, and processes of wider political engagement (Iriye, 2017). These include what a growing number of scholars in multiple fields see as the rise of cities in international diplomacy and planetary policy making (Engelke, 2013). In other sense, however, the dominant language and interpretative frameworks of globalization and anti-globalization—these political dualities especially—have grown stale. The analytical advantages of re-writing through traditions of internationalism, albeit not without its own problems as an organizing narrative, relate to long-standing multiplicities and contestations (Sluga & Clavin, 2017b).

Historians like Mark Mazower, Akira Iriye, Glenda Sluga, and Patricia Clavin have recently revived these multiplicities and contestations—pointing out that, like a fast river that hits a wide delta, internationalism almost immediately split into many internationalisms. Parallel, often entangled, sometimes merely ephemeral, developments carved out different kinds of geopolitical landscapes. Some groups followed (and still follow) the channels opened by classical economists like Adam Smith and David Ricardo, though more defensively these days. The way to a peaceful and harmonious world order, they reasoned, was through free trade rather than neo-mercantilist economic nationalism (or what is today often called economic populism). Leading advocates like Richard Cobden, the industrial entrepreneur from the shock city of Manchester and the man arguably most responsible for parliamentarian repeal of the British Corn Laws in 1849, successfully advocated both at home and abroad for tariff reforms, the reduction of protective trade barriers, and the peaceful negotiation of commercial treaties. The Corn Laws originally protected rural producers in the first half of the nineteenth century. Their demise signaled a shift in political (and electoral) power to urban-industrialized populations.

This new liberal internationalism of the free trade movement—popularly championed without fail by *The Economist* since 1843—emerged from the early and deep industrialization of British social space, suggesting critical yet too often neglected links between new kinds of urban geographies and new kinds of global-historical politics and attendant discourses. In the post-World War II twentieth century, as Clavin (2017) further argues, the liberal river gathered speed—carving out more territory as the *agency* of central bankers established novel rules for new rounds of public investment and private accumulation.

Yet integrated and uneven markets shaped only so much ground. Social justice and moral challenges motivated many other groups. Issues like slavery, world peace, and the plight of women and children tapped into the fervor of evangelical values and Christian ethics of spiritual responsibilities. These values and ethics sometimes overlapped with the liberal view, as in the Peace Movement and early forms of humanitarianism (Clavin, 2017). But for Mazower (2012), justice movements sought to forge new conceptions of "brotherhood" around alternative kinds of social identities. As they battled each other for influence in the working-class halls of Victorian London, the Italian nationalist, Giuseppe Mazzini, and the communist activist and social theorist, Karl Marx, offered still other conceptions of international bonding. Left "cold" by Cobden's efforts "to deify self-interest" (ibid.), Mazzini saw a future built around the self-determination of culturally coherent nations and states. Marx, of course, thought that a new class consciousness would eventually lead to a Socialist International of revolutionary workers—and eventually to the utopia of post-capitalist communism. In the twentieth century, Mao's peasant armies carried the internationalist banners of The Great Helmsman—but also of Stalin and Lenin.

Interestingly, lawyers and scientists conceptualized and activated their own brands of internationalism. The legal profession placed emphasis on codification schemes, arbitration, and appropriate rules of global engagement—in the bloody business of war and the commercial war of business. Statisticians, engineers, geographers, public health experts, etc. from different countries met increasingly "to push aside well-connected amateurs and bring in new cadres of educated and rational elites" (p. 95). European cities like London, Brussels, The Hague, and Geneva now gained a novel role in history: They became international political platforms for ideas, data, and agreements that steadily nourished what H.G. Wells would later call "the World Brain" (p. 110).

A young soldier standing in the muddy trenches of Passchendaele in 1917 or the burning wreckage of Dresden, Germany thirty years later could dismiss these variants of internationalism as naive—even abjectly cruel. World peace? Christian comity? Rationality? Socialist brotherhood? Free trade? Technological progress? Rules for war? Where once Victor Hugo could enthusiastically claim that "Everything is moving at once—political economy, science, industry, philosophy, legislation" (p. 35), the discordant poetry of Siegfried Sassoon, the collapse of the Great Depression, Stalin's gulag, Hitler's genocide, the split atom, and even (in time) the less dramatic but steady ecological violence of mass production economies suggested a far darker reality. Lewis Mumford, the American urbanist, thought the foremost impact of 1914–1918 on his generation was "the loss of to-morrow" (cited in Miller, 1989, p. 164). The roar of the twenties masked in jazz riffs the aching pain of innocence lost. The iconic architecture of the human experience in August, 1945 now seemed, alas, the death camp (Fig. 2.1). The first half of the twentieth century—so unlike the long Victorian period—seemed to shatter

Fig. 2.1 The Nazi death camp in 1945 (*Credit* Everett Nitzke; author's personal collection)

into tiny shards the available international heritage and with it the possibility of a global political commons.

Remarkably, it did not. Many kinds of twentieth-century internationalisms soon thrived again—from renewed concerns with religion to new movements around feminism, human rights, indigenous culture, and geostrategic calls for a New International Economic Order (NIEO) in the 1970s by non-aligned players in Global South, notably India (Sluga & Clavin, 2017b). None attained hegemonic status. None displaced the traditional role of nation-states and the Westphalian interstate system. But they were each important in shifting how we conceive of the politics, politics, and programs of global interactions in the early twenty-first century. The promotion of international human rights immediately after World War II, for instance, was still overwhelmingly state-centric and juridical in nature. Yet by the 1970s, as Roland Burke (2017, p. 288) shows, "NGO activism adopted a vision of rights defined by parsimony, advanced by social movements that were typically distant from national political levers"; by the 1990s, he further shows, the movement broke from statism in ways that now reflected, as never before, a wider post-Westphalian world order, about which more throughout this book.

At the same time, the post-World War II era saw the attempted, globally hegemonic, reconsolidation of a managed liberal world order under the security umbrella of the USA. In truth, this was hardly the stripped-down international project propounded by the Cobdian free traders a century earlier. Within Western Europe, interesting new forms of social democracy channeled the raw imperatives of markets into new kinds of geopolitical economies, particularly in the small, more homogenous, Nordic countries (Swenson, 2002). But even within the more socially diverse USA, where race and ethnicity had long militated against a more progressive model (Lipsett & Marks, 2000), a significant federalization of social, economic, and environmental policies during the New Deal blended together liberalism with other kinds of values and ideologies, especially social planning rationalities (Maher, 2008).

Framed by post-war geopolitics, the American New Deal culture and the experiences of wartime military planning invariably seeped into international space as the world community focused on physical reconstruction, societal recovery, and economic expansion. The Bretton Woods system created the International Monetary Fund (IMF), the General Agreement on Trade and Tariffs (GATT), and the World Bank. These institutions helped to stabilize the international monetary system around fixed exchange rates based on gold reserves. Bretton Woods promoted freer trade between nations and especially the reduction of tariffs. But this required government intervention within and across sovereign national borders as well as the tighter and better coordinated public management of private economies. Moreover, the architecture of the United Nations, including new bodies like UNESCO (focusing on science, education, and culture), WHO (health), and FAO (food and agriculture), created a now legible supranational "layer" of formal international institutions far more capable, at least in theory, of both studying and to some extent promoting integrated norms around science, policies, and especially "development" thinking. This interstate conceptualization of global order drew on the market-liberal, legal-institutional, and scientific-technocratic variants of internationalism—but was based paradoxically on the hegemonic power of the liberal USA.

The liberal order, like Westphalia, was always something of a Great Unicorn—a normative aspiration as much as an actual reality. Just as the Holy Roman Empire was neither Holy nor Roman nor really much of an Empire, the Liberal International Order was only partly liberal, incompletely international, and far from ordered. For one thing, freedom was never free, and free trade was hardly fair trade, as the NIEO movement insisted in the 1970s. Market liberalism required heavy state steering and firm institutional controls, leading to what Edward Luttwark eventually called a system of "controlled capitalism" that invariably reflected the power dynamics of strong nations and of the elite groups within those nations (cited in Steger, 2002, p. 26). Other competing models of political economy—the Soviet and Chinese cases, above all, but also authoritarian and localized aberrations like South African apartheid and Portuguese Salazarism—further suggested ongoing variety across international political space throughout the twentieth century.

For a moment in the early 1990s, however, the liberal unicorn seemed almost real to some believers. Faith in liberal democracy was now peaking. The collapse of Soviet-style socialism; the economic opening of China in the 1980s; the end of the Cold War in 1990; the final death of apartheid in 1994—all this suggested what Francis Fukuyama (1992) famously theorized as "the end of history," by which he meant the ideological, geopolitical, and especially intellectual-normative triumph of liberal democracy as the only plausibly effective model of political economy left standing. From only three (thin) democracies in 1790, Fukuyama counted about thirty in 1975, but over sixty by 1990. For him, the ideological and intellectual struggle was over. The logical conclusion to make from such trends was obvious. Fukuyama's thesis, albeit more nuanced than his many critics typically allow, is bereft of any postmodern angst about hegemonic meta-narratives. "There is a fundamental process at work that dictates a common evolutionary pattern for *all* human societies," he suggested; "—in short, something like a Universal History of mankind [sic] in the direction of liberal democracy" (p. 48).

Alas, no. It was less the triumph of liberal democracy that seemed especially important to many observers of the 1990s than something altogether different: the corporate consolidation of neoliberal capitalism, a process that exaggerated economic inequalities politically dangerous to the liberal theory of society. While part of this consolidation included the "roll back" of the twentieth century Keynesian welfare state (Peck, 2008), it is perhaps more accurate to see, with George Monbiot (2000) and others, an equally important process of "state capture," a reality distinct from the classical liberal project of the nineteenth century. The core distinction between liberal and neoliberal political economies is illustrated simplistically in Fig. 2.2.

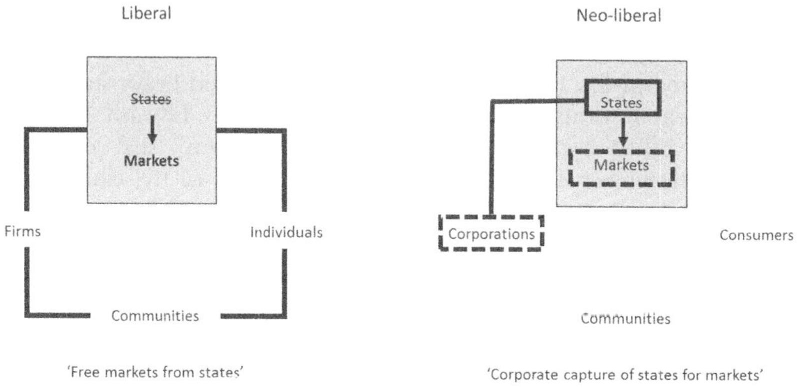

Fig. 2.2 State capture: liberal versus neoliberal political-economies

Classical liberalism, in this view, seeks to free markets from states; neoliberalism, in contrast, seeks to reconstruct markets through a captured state, sidelining or trying to make politically irrelevant the collective policy preferences of various communities while transforming individuals into consumers through systemic modes defined disproportionately by corporations. For instance, neoliberal rationalities have helped governments to turn "citizens" into "customers," all while reconstructing fresh spaces of accumulation safe for capital. A huge stream of comparative research on neoliberal urbanism has documented the variegated geographies and politics of these changes (Bond, 2000; Didier, Morange, & Peyroux, 2013; Hackworth, 2007; Peck, 2008). Jason Hackworth (2007) writes of "the neoliberal city" as an urban type, contrasting this with, for example, Pierre Clavel's (1986) earlier planning work on "the progressive city." Ultimately, the neoliberal project ensures for its many critics that the biggest corporations do not really compete at all. They practice a distinctive type of "alliance capitalism" (Brennetot, 2015; George, 1999; Gunder, 2010; Hackworth, 2007).

Within this historical context, work on the *geographies* of internationalism tend to reflect a growing concern with different forms of "subaltern" resistance to the hegemonic neoliberalization of state and society—and thus to the corporatization of cities (Featherstone, 2015). As might be expected, much if not all of this critical geographical scholarship reengages with labor movements and the transnational and supra-local solidarities of work and production (Ferretti, 2018; Glassman, 2009; Waterman & Wills, 2001).

One of the most interesting contributions to the body of work on labor internationalism includes Franco Barchiesi's (2001) analysis of the cross-border solidarity forged by the South African Municipal Workers Union (SAMWU). Barchiesi highlights the limitations of SAMWU's internationalist strategies after the end of apartheid in the early 1990s, but nonetheless believes that "the struggle for decommodifation of basic services provides a potentially important terrain for mobilization in which trade unions could reinvent themselves in representing their members' as workers and as citizens, and relate to global social movements on this basis" (pp. 98–99). In other words, organized labor in South Africa, as elsewhere, has had to work harder to find new global solidarities *beyond* organized labor; this creates organizational and material challenges. "Old" forms of labor internationalism familiar to nineteenth and twentieth century politics, as hard as that was/is in the face of organized capital, now have to give way to "new" internationalist terrains populated by wider linkages with, inter alia, women's struggles; intellectuals and students defending the public academy; faith-based activists; indigenous rights campaigns; and finally environmentalists resisting the constant commodification of natural resources through the privatization of communal spaces (p. 83). This cosmopolitan and more complex vision of internationalism is closer to Hardt and Negri's (2000) theory of a "multitude" challenging the "empire" of global capital. For Waterman and Wills (2012, p. 2), such a vision means getting "beyond the fragments" of a trade union tradition "…ill-suited to the coalition politics and alliance-building so critical to the contemporary countercultural project."

Largely because the territorial state is considered implicated and comprised, radical geographers of internationalism have accordingly focused more on non-state, society-based manifestations of trans-border changes in the social relations of alliance-building, notably labor and associated groups. Yet considerable work within political and urban geography has explored what Neil Brenner and others have long called the "new state spaces" produced by recent shifts in the economics and politics of global and local capitalism—and thus the parallel crises of ecological resiliency and uneven social development (Brenner, 2004; Curtis, 2016; Herrschel, 2013; Herrschel & Newman, 2017; Moisio, 2018). As briefly mentioned Chapter 1 but worth reiterating here, Andy Jonas and Sami Moisio, both separately and together, have linked the recent rise of larger city-regions—and what they and others call the emergence of city-regionalism (Jonas, 2015)—to a *new* kind of global geopolitics. These global geopolitics—these distinctive, new kinds of internationalism, as I am instead calling it in this book—retain the state as a central actor in the overall story (Jonas, 2012).

But again, that same state is changing forms, reassembling, morphing, and shifting its scalar and territorial configurations (Curtis, 2016). The *urbanization* of the state, Moisio (2018) shows, is an important development in this regard. Anne-Marie Slaughter (1997) first noted, for instance, that the original promise of the liberal world order relied on centralized rule-making authority and universal membership, which the United Nations could neither provide nor ensure. Some scholars in the 1990s, such as Jessica Mathews (1997), thus predicted the counter-emergence of a "new medievalism," wherein power would steadily shift away from national states—up, down, and sideways. Slaughter instead emphasized "transgovernmentalism [as] the most widespread and effective mode of international governance." She posited over twenty years ago,

> [a] new world *is* emerging, with less fanfare but more substance than either the liberal internationalism or new medievalist versions. The state is not disappearing, it is disaggregating into its separate, functionally distinct parts [, … which] are networking with their counterparts abroad, creating a dense web of relations that constitute a new, transgovernmental order.

For Slaughter, "the densest area" of this hypothesized web of relations increasingly involved not (yet) cities but national regulators charged with more mundane but still influential arenas of global politics. Common examples included—and still include—antitrust enforcement, securities regulation, banking and insurance supervision, criminal law enforcement and, of special interest to us here, what she generically called "environmental policy." Paradoxically, it may be that the limitations of national regulators in green policies created new spaces for cities to act.

Slaughter's ideas complement Mazower's themes around early forms of legal and scientific internationalism. Similarly, Barchiesi's work picks up the evolution of labor internationalism. Both suggest, as I do here, a longer running and more complex, punctuated story with multiple possibilities. Slaughter's conception of transgovernmentalism is, to be sure, a novel kind of inter-*national*-ism; so too are the theoretical and practical efforts of a broadened laborite internationalism that more effectively links across increasingly diverse social movements—even as political tensions between organized labor and environmentalist groups should not be underestimated (Clapp & Dauvergne, 2008). (The grassroots is a complex ecosystem, too, and building alliances across places and interests is arguably more challenging than integrating corporate offices across global space.) Yet both themes look less original when placed in the long run of global political history. Both themes are part of how multiple orders have dialectically influenced each other in global space, at least since the nineteenth century and perhaps longer.

In the next section, I develop this last idea a bit further, ultimately suggesting we need a wider "frame" that helps us to move beyond the classic Westphalian project of state-to-state diplomacy that has for so long dominated our imaginaries of global politics and cross-border collective action. Many observers now routinely highlight the growing limits and even "disintegration" and "crisis" of the long-familiar Westphalian world order (Newman, 2009). Carlo Bondoni (2013) argues, for instance, that economic globalization has steadily undermined the political rules of international relations, eroding legal and cultural boundaries between states. Yet the hypothesized shift to a post-Westphalian world, I again want to argue, is less about *replacement* than *adaptation*; about the abrasion and imposition of modern-national and post-modern-cosmopolitan worlds, and therefore about scalar, spatial, and political co-relationships between cities, states, and global environmental problems. It is not easy to grasp the geographies of these relationships; but as the book unfolds I shall attempt to sketch out some of the likely many urban spatialities arguably reconfiguring the green shape of world affairs.

Reimaging Global Political Order(s): Toward a Wider Frame?

The dominant frame through which we see global political space—the frame that traps us into a cartography of countries interacting with one another through territorial states—is central to the Westphalian world order that emerged after the famous Peace of Westphalia in 1648. The details of this treaty are less important here than the three main theoretical principles that have defined this order, *once again*: (1) state sovereignty; (2) legal equality of states; and (3) the non-intervention of one state in the affairs of another. While the violation of these normative principles through colonialism, imperialism, warfare, and neocolonial economic hegemony are no less common than their protection, Westphalian global space nonetheless represents for many scholars a key innovation in world political development (Hayman & Williams, 2006).

Westphalian norms (and practices) in world politics have been part and parcel of the larger conceptualization of a specific kind of modernity—involving the displacement of religious order and cosmology in the conduct of public affairs at multiple scales of power with rationality, science, and especially the apparatus of territorial states that, with the eventual demise of empires, theoretically and substantively comports with nations. Put another way, national interests were—and are—"carried into" global space by differently endowed territorial states who in turn *expect to* enjoy sovereignty within agreed borders (Amen et al., 2012). So imagined, internationalism is the world political space forged by the interstate system (Fig. 2.3).

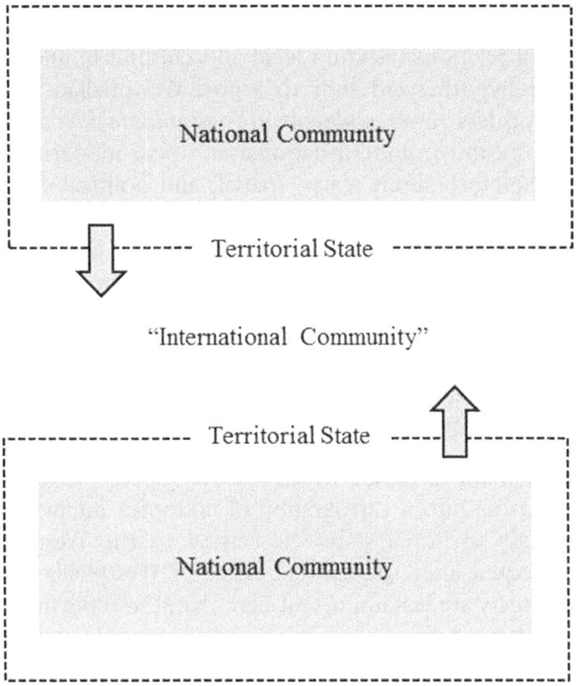

Fig. 2.3 The Westphalian "frame": internationalism *through* the state

Still, nations and states are not separate developments. Nations and states co-constitute each other dialectically over time. The rise of the territorial state concomitantly helped and depended upon the congealing of national identities—or national communities. American national identity, for example, did not *precede* the American state. It was and is an ongoing project of that state (e.g., millions of public schoolchildren pledge allegiance to the American flag every single morning). State-building and nation-building, in the USA as elsewhere, are deeply imbricated. But they are different things. "State and nation go together and support each other," Carlo Bodoni (2013) notes. Shifts in one means shifts in the other. A crisis emanating in the state apparatus has implications for the constitution of national communities—and vice versa.

Paradoxically, the theoretical power of the Westphalian frame to capture international relations has been as much normative as positive. As Fukuyama (2013) has argued, very few societies today are like Denmark, where state and nation essentially "match up" and thus, for this reason, supposedly function more effectively than elsewhere in the world. But "getting to Denmark" nonetheless has been an important political standard—a desired norm—not least around how national communities are supposed to establish effective local political order, in general, and global sustainability, in particular. This Westphalian "frame" for how world politics works suggests that internationalism is conducted through the territorial state.

We know better. Or at least we know that this particular frame is losing its power to provide workable norms much less real explanations. It filters out too much of what has been actually happening across the world over the past several decades—and in fact much longer. Globalization, for one thing, has steadily enervated the territorial state's historic (if always slightly mythical and reversable) capacity to excerise Westphalian-ordained sovereignty over "its" peoples, economies, and societies. States struggle mightily with flows—and flows, while always part of human history, have exploded and intensified in recent decades. In addition, new "imagined communities," as Benedict Anderson (2016) and Arjun Appadurai (1996) both famously suggested, have formed and gradually mutated, vastly complicating political narratives which too rigidly funnel the whole of global space into singularly congealed nations through "state-to-state" internationalism. As a consequence, despite the ongoing importance of states today we really do need a much wider "frame" (Fig. 2.4).

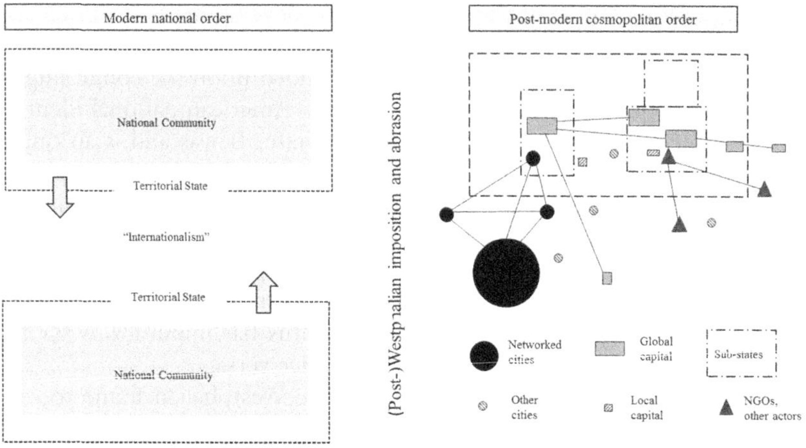

Fig. 2.4 (Post-)Westphalian imposition and abrasion

That modified frame should not *efface* the old frame *tout court*. It should consider how the old frame might better represent new realities—or indeed might better capture older realities that have been ignored, downplayed, and/or under-appreciated all along, including the many traditions of internationalism just discussed. Hence the post-Westphalian frame depicted does not suggest that the contemporary world has moved totally beyond the Westphalian politics of territorial states and their international organizations, including the United Nations or the World Trade Organization. In reality, there is a strong case that such politics remain geopolitically dominant, not least in classic arenas of international politics such as military alliances and defence agreements. At least so far, after all, only states control nuclear weapons. Whether or not and how state-based dominance is part of our problem is taken up in the next chapter. For now, though, we need only consider that, rather like the dialectics of nation and state, Westphalian and post-Westphalian processes have *similarly constituted one another* in various ways over time and space.

The shift to a post-Westphalian frame recognizes these dynamics. First and most importantly, the post-Westphalian frame recognizes that international actors are not always (or only) states carrying out national interests. State departments and foreign ministries are certainly important but hardly act alone—nor have they for some time. This theme was actually developed by Robert Keohane and Joseph Nye (1973) in the early 1970s and was already evident earlier in history (Amen et al., 2012). Put in a slightly different way, the so-called international arena, where global (geo)politics and struggles all play out, is populated increasingly by non-state actors advancing a range of non-state interests and values not readily captured by their respective national communities or origins. As already discussed in Chapter 1 and earlier in this chapter, for example, but worth repeating here, many scholars now see cities as key actors in international relations (Calder & Freytas, 2009; Curtis, 2014), about which more in Chapter 4.

A second dynamic recognizes the internal complexity of the state. As Amen et al. (2012) observe in their important contribution, Morton Halperin's (1974) *Bureaucratic Politics and Foreign Policy* showed how states are not really unitary actors. They are assemblages of sometimes complementary, often competing, institutions. In his discussion of "state-building," Fukuyama similarly distinguishes various state functions along two axes: scope and strength. He argues (albeit conservatively) that states should limit their "scope" of functions to core institutional competencies like defense, law and order, and macroeconomic management, but also, neo-conservatively, that they should strengthen their institutional capacities to carry out these functions as intended. That is what he means by "state-building," a crucial problem in global politics.

For Fukuyama, the world order is (or was until recently) composed of four main kinds of internally complex states: Quadrant 1 states who seek to do *few* things and do them reasonably well (e.g., New Zealand); Quadrant II states who seek to do *many* things and do them reasonably well (e.g., France); Quadrant III states who seek to do *few* things, but do them *poorly* (e.g., Sierra Leone); and Quadrant IV states who try to do *many* things, but do them *poorly* (e.g., Brazil). Global state history for him is about mapping "movements" within and across these four spaces, as shown in his illustration of political change over time that I reproduce in Fig. 2.5.

Notwithstanding this representation of stateness, Fukuyama, as might be expected, detected an unevenly experienced yet overall, i.e., worldwide, movement away from neo-Keynesian social democratic aspirations toward more liberal formations of a limited state. In other words, all the arrows were moving "to the left" on the diagram below, suggesting a steady retrenchment of the territorial state's overall *scope* of functions, including "intermediate" functions like environmental protection and

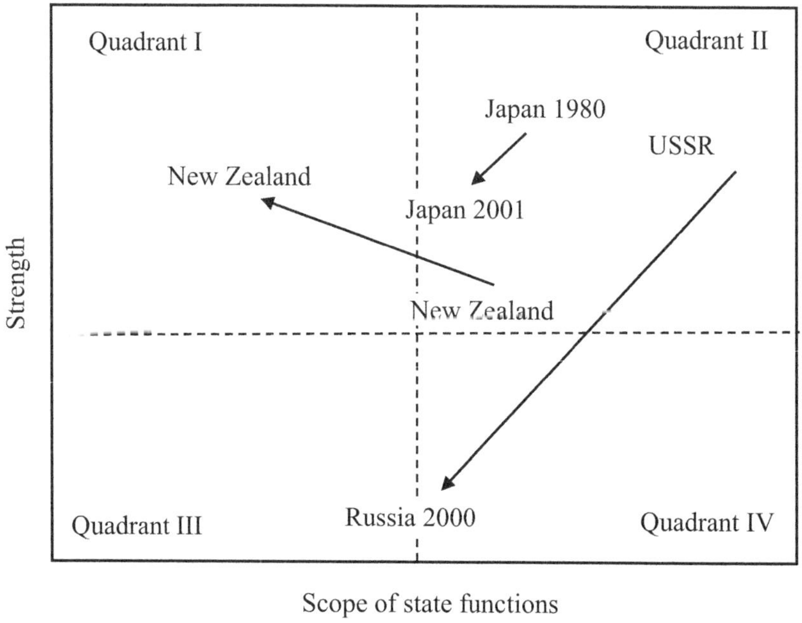

Fig. 2.5 State-building and political development (Fukuyama, 2004)

"activist" functions like fostering markets. In reality, the liberal project of recent decades has metastasized into a corporate neoliberal project, wherein territorial and socioeconomies inequalities have generated an anti-liberal, populist backlash.

This paradoxically may have opened up new room for cities. While Halperin's core insights reflect neo-Weberian sensitivities to complex state formation (Skowronek, 1982) and global policy projections (Seabrooke, 2002), they have similarly characterized work by scholars of urban political economy interested in global politics and international problems. In his comparative study of cities, *Governing from below*, for instance, Jeff Sellers (2002) similarly noted that "actual modern states encompass dozens of institutionally distinct policy sectors with highly diverse organizational architectures." In addition, he suggested, "[c]onfigurations of state policies, institutions, and actors may assemble in very different ways in one region or locality than in another. [In other words] local and regional regimes of territorial governance…*differ widely among places even under the same matrix of national policies and institutions*" (emphasis added). Because the state is "polycentric" so are its political geographies and socio-scalar projects (Morgan, 2007).

A third dynamic recognizes how the rise of cities as international actors does not necessarily resolve the complex polycentricity of the state at the urban scale. As an emerging literature on urban politics and social-spatial development policy argues, cities are not unitary "things" either; they do not "act" the way a person acts. Cities are assemblages of multiple institutions and forces and therefore are subject to what Orren and Skowronek (2004) call "intercurrence," defined by others as "the coexistence of multiple orders, typically originating at different times and in tension with one another…" at/in any given site (Stone & Whelan, 2009, p. 99). One major consequence follows from this core claim:

> The very tendency of institutions to persist means that at any moment in time several different sets of rules and norms are likely to be operating simultaneously. To the extent that the idea of order presumes institutions synchronized with one another, entailing their creation all at once, something unlikely to be accomplished by even the most radical revolution. These insights take on special significance in the case of political institutions, because political institutions are inherently other-directed; that is, they seek to control individuals or other institutions outside their own sphere. As a consequence, different institutional rules and norms will abut and grate as a normal state of affai*rs*. (Orren & Skowronek, 1996, p. 112)

While these ideas were developed originally to describe and explain American Political Development (or "APD"), scholars have increasingly both *urbanized* and "de-Americanized" APD's key theoretical concepts, especially intercurrence, which strikes many international observers as conceptually expansive enough to apply productively to diverse scalar and territorial contexts (Dierwechter, 2017; Dilworth, 2009; Hodos, 2009; Lucas, 2015).

The danger, it must be said, is that we take these propositions too far, whether normatively or analytically. We can and should insist, for example, on the global importance of "non-state" actors, on the transnational agency of various "communities within nations," including new urban political "communities" like C40 or City Protocol who are each acting in legitimately novel ways to take on global problems like climate change. We can and should pay careful attention to the polycentricity of the territorial state apparatus, even as we are just as careful to apply this same polycentricity to cities. But we should not assume that Westphalian forms of global politics emphasizing the role of territorial states have vanished; that states do not matter anymore; that state institutions do not have important effects on post-Westphalian global politics and possibilities.

In sum, we should not see post-Westphalian internationalism as “post-state.” We should see a more complex intermingling of political forces and geographical projects—demanding a widening of our overall interpretive frame. Post-Westphalian politics dialectically incorporates and modifies Westphalian spatialities; they do not replace them. The world suffers from failing states; yet it is not so much “the state that is failing,” as Simon Curtis argues, “but a particular historical iteration of the state – the nation-state – that is being reassembled into a new form” (op cit.) As historical iterations change, it logically follows, geographical iterations change as well. Yet our urban atlas of these new global spaces is by and large empty.

Conclusions

Internationalism challenges how we see the operation of urban policies. When most of us close our eyes and imagine the organizing geography of global space, we still do so through a strong conceptual frame that actively partitions and cleanly segments the planet’s physical continents. We see tightly bordered nations governed by strongly territorialized states. We see global political space the way we see a jigsaw puzzle. We snap in Italy and Zambia here; Bolivia and Nepal there—until our world is cartographically full. When most of us close their eyes and imagine the organizing meta-geography of global space, in a word, we see the Westphalian world order of sovereign states representing cohesive nations as they interact with one another.

We know better. We are aware, as a first problem, of important exceptions and necessary caveats. Some nations, like the Kurds, have no corresponding states; they are governed by multiple states extending various rights and extracting various obligations. Other states, like Somalia and Yemen, are either fragile or failing—flags in search of countries. Still other states, including Spain, Canada, India, and the UK, periodically and in some cases constantly negotiate the power dynamics of multi-national identities and attendant territorial-regionalist cultural and economic claims. These and a host of related issues constitute much of what is most important and far-reaching in international affairs today.

But there are more than just "mismatch" issues to consider. A generation ago, John Agnew (1994) argued that our global political imaginary, and hence a great deal of work on world politics, had long suffered from what he memorably called the "territorial trap." The implicit conceptual and empirical assumption usually has been, Agnew claimed, "that a state is a fixed territorial entity (even if its actual boundaries can change) operating much the same over time and irrespective of its place within the global geopolitical order; a state is territorial much like life on earth is terrestrial" (p. 54). The power of this trap in the present century is that it helps us conceptually to carry imagined national communities into international political space through essentially homogenous state structures. But as just argued, states are neither singular objects, nor built all at once, nor working alone across national boundaries—today or in the past. For these reasons, they are unlikely to work alone in the future.

The emerging histories and geographies of internationalism discussed in this chapter have opened up and expanded upon long-neglected stories that complicate extant readings of global politics, including the urbanized politics of the global environment that are discussed in the next two chapters. Past patterns in variegated, often contending brotherhoods—labor, scientific, moral-ethical, legal, nationalist, etc.—have both informed and steadily evolved over time into current concerns with a multitude of cross-border, often subaltern concerns, including how to engage geopolitically with globally shared problems.

The selective rise of networked cities now focused on global challenges like climate change and smart city transitions, discussed later in Chapter 4, is in *some* ways novel as a distinctive form of internationalism. Yet it is also part of a much wider terrain of interlinked continuity and change that suggests a historically complex, geographically striated, and politically open set of international outlooks and practices. We should resist a periodization that suggests clean passages from state to non-state; from national to urban; from Westphalian to post-Westphalian. David Bensaid's theoretical sense that a plurality of temporalities is still playing out across political space helps us to capture the scalar "sliding" up, down, and sideways both within and outside the formal (inter)state system. Bensaid helps us to understand the internationalization of cities, the urbanization of internationalism, and the polycentricity of a changing, adapting, state system as it engages with cities and other actors "below" and "above" and "across" its traditional territorial moorings. The rise of cities in global affairs reflects the reassembling of the modern state no less than the ongoing importance of non-state actors in forging cosmopolitan outlooks and practices.

In short, a jigsaw puzzle map of the world and its political orders will no longer do—if indeed it ever did. Nowhere is this clearer than in how various actors have engaged with global ecological problems, the synoptic theme to which I now turn.

References

Agnew, J. (1994). The territorial trap: The geographical assumptions of international relations theory. *Review of International Political Economy, 1*(1), 53–80. https://doi.org/10.2307/4177090.

Agnew, J. (2000). Global political geography beyond geopolitics. *International Studies Review, 2*(1), 91–99.

Amen, M., Toly, N., McCarney, P., & Segbers, K. (Eds.). (2012). *Cities and global governance: New sites for international relations.* Farnham, UK: Ashgate.

Anderson, B. (2016). *Imagined communities : Reflections on the origin and spread of nationalism.* London: Verso.

Antentas, J. (2015). Sliding scale of spaces and dilemmas of internationalism. *Antipode, 47*(5), 1101–1120. https://doi.org/10.1111/anti.12169.

Antentas, J. (2016). Daniel Bensaïd, Melancholic strategist. *Historical Materialism, 24*(4), 51–106. https://doi.org/10.1163/1569206x-12341500.

Appadurai, A. (1996). *Modernity at large cultural dimensions of globalization.* Minneapolis: University of Minnesota Press.

Attwell, W. (2014). The rise of cities as global actors: What consequences for policy? *Global Policy, 5*(3), 374–376.

Barchiesi, F. (2001). Transnational capital, urban globalisation and cross-border solidarity: The case of the South African municipal workers. In P. Waterman & J. Wills (Eds.), *Place, space and the new labour internationalisms.* Oxford, UK; Malden, MA: Blackwell.

Bodoni, C. (2013, February 12). *A crisis of the state? The end of the postWestphalian model.* Retrieved from http://www.socialeurope.eu/author/carlo-bordoni/.

Bond, P. (2000). *Elite transition: From apartheid to neoliberalism.* London: Pluto Press.

Brenner, N. (2004). Urban governance and the production of new state spaces in Western Europe, 1960–2000. *Review of International Political Economy, 11*(3), 447–488. https://doi.org/10.2307/4177507.

Brenner, N. (2014). *Implosions/explosions: Towards a study of planetary urbanization.* Berlin: Jovis.

Brennetot, A. (2015). The geographical and ethical origins of neoliberalism: The Walter Lippmann Colloquium and the foundations of a new geopolitical order. *Political Geography, 49*, 30–39. https://doi.org/10.1016/j.polgeo.2015.06.007.

Brooks, C. (2015). The apostle of internationalism: Stephen Duggan and the geopolitics of international education. *Political Geography, 49*, 64–73. https://doi.org/10.1016/j.polgeo.2015.05.007.

Burke, R. (2017). The internationalism of human rights. In G. Sluga & P. Clavin (Eds.), *Internationalisms: A twentieth-century history*. Cambridge, UK: Cambridge University Press.

Calder, K., & Freytas, M. (2009). Global political cities as actors in twenty-first century international affairs. *SAIS Review of International Affairs, 29*(2), 79.

Chatterton, P., Featherstone, D., & Routledge, P. (2013). Articulating climate justice in Copenhagen: Antagonism, the commons, and solidarity. *Antipode, 45*(3), 602–620. https://doi.org/10.1111/j.1467-8330.2012.01025.x.

Clapp, J., & Dauvergne, P. (2008). *Paths to a green world: The political economy of the global environment*. Cambridge, MA: MIT Press.

Clavel, P. (1986). *The progressive city*. New Brunswick: Rutgers University Press.

Clavin, P. (2017). Men and markets: Global capital and the international economy. In G. Sluga & P. Clavel (Eds.), *Internationalisms: A twentieth-century history*. Cambridge, UK: Cambridge University Press.

Curtis, S. (Ed.). (2014). *The power of cities in international relations*. New York: Routledge.

Curtis, S. (2016). Cities and global governance: State failure or a new global order? *Millennium, 44*(3), 455–477. https://doi.org/10.1177/0305829816637233.

Czarniawska-Joerges, B. (2002). *A tale of three cities: Or the globalization of city management*. Oxford: Oxford University Press.

Didier, S., Morange, M., & Peyroux, E. (2013). The adaptative nature of neoliberalism at the local scale: Fifteen years of city improvement districts in Cape Town and Johannesburg. *Antipode, 45*(1), 121–139. https://doi.org/10.1111/j.1467-8330.2012.00987.x.

Dierwechter, Y. (2017). *Urban sustainability through smart growth: Greater Seattle, intercurrence, and the new geography of regional development*. New York: Springer.

Dilworth, R. (Ed.). (2009). *The city in American political development*. New York: Routledge.

Engelke, P. (2013). *Foreign policy for an urban world: Global governance and the rise of cities*. Washington, DC: Atlantic Council.

Featherstone, D. (2012). *Solidarity: Hidden histories and geographies of internationalism*. London and New York: Zed Books.

Featherstone, D. (2015). Maritime labour and subaltern geographies of internationalism: Black internationalist seafarers' organising in the inter-war period. *Political Geography, 49*, 7–16. https://doi.org/10.1016/j.polgeo.2015.08.004.
Ferretti, F. (2018). Geographies of internationalism: Radical development and critical geopolitics from the Northeast of Brazil. *Political Geography, 63*, 10–19. https://doi.org/10.1016/j.polgeo.2017.11.004.
Forman, M. (1998). *Nationalism and the international labor movement: The idea of the nation in socialist and anarchist theory*. University Park, PA: Pennsylvania State University Press.
Frankowski, P. (2010). One world-many orders? In S. Krishna-Hensel (Ed.), *Order and disorder in the international system*. Farnham: Taylor and Francis.
Fukuyama, F. (1992). *The end of history and the last man*. New York: Maxwell Macmillan International.
Fukuyama, F. (2004). *State-building: Governance and world order in the 21st century*. Ithaca, NY: Cornell University Press.
Fukuyama, F. (2013). *Political order and political decay*. New York: Farrar, Strauss & Giroux.
George, S. (1999). *A short history of neoliberalism: Twenty years of elite economics and emerging opportunities for structural change*. Paper presented at the Conference on Economic Sovereignty in a Globalizing World, Bangkok.
Glassman, J. (2009). Critical geography I: The question of internationalism. *Progress in Human Geography, 33*(5), 685–692. https://doi.org/10.1177/0309132508101602.
Gunder, M. (2010). Planning as the ideology of (neoliberal) space. *Planning Theory, 9*(4), 298–314.
Hackworth, J. (2007). *The neoliberal city: Governance, ideology, and development in American urbanism*. Ithaca: Cornell University Press.
Halperin, M. (1974). *Bureaucratic politics and foreign policy*. Washington, DC: Brookings.
Hardt, H., & Negri, A. (2000). *Empire*. Cambridge, MA: Harvard University Press.
Hayman, P., & Williams, J. (2006). Westphalian sovereignty: Rights, intervention, meaning and context. *Global Society, 20*(4), 521–541. https://doi.org/10.1080/13600820600929879.
Herod, A. (2004). Geographies of labor internationalism. *Social Science History, 27*(4), 501–523. https://doi.org/10.1017/s0145553200012669.
Herrschel, T. (2013). *Cities, state and globalisation. City-regional governance in Europe and North America*. London: Routledge.
Herrschel, T., & Newman, P. (2017). *Cities as international actors: Urban and regional governance beyond the nation state*. London and New York: Palgrave.
Hodder, J., Legg, S., & Heffernan, M. (2015). Introduction: Historical geographies of internationalism, 1900–1950. *Political Geography, 49*, 1–6. https://doi.org/10.1016/j.polgeo.2015.09.005.

Hodos, J. (2009). Against expectionalism: Intercurrence and intergovernmental relations in Britain and the United States. In R. Dilworth (Ed.), *The city in American political development* (pp. 44–63). New York: Routledge.

Imlay, T. (2017). Socialist internationalism after 1914. In G. Sluga & P. Clavin (Eds.), *Internationalisms: A twentieth-century history* (pp. 213–242). Cambridge, UK: Cambridge University Press.

Iriye, A. (2017). Forward. In G. Sluga & P. Clavin (Eds.), *Internationalisms: A twentieth-century history*. Cambridge, UK: Cambridge University Press.

Jonas, A. (2012). City-regionalism as a contingent 'geopolitics of capitalism'. *Geopolitics, 18*(2), 284–298.

Jonas, A. (2015, October 8–9). *Variegated city regionalisms: The geopolitics of city-region building*. Paper presented at the Smart City-Regional Governance, Cosmopolis, Centre for Urban Policy Research, Vrije Universiteit Brussel, Brussels.

Keohane, R. O., & Nye, J. S. (1973). Power and interdependence. *Survival, 15*(4), 158–165. https://doi.org/10.1080/00396337308441409.

Krishna-Hensel, S. (2010). *Order and disorder in the international system*. Farnham: Taylor and Francis.

Lefebvre, H. (2013). *Rhythmanalysis: Space, time, and everyday life*. New York: Bloomsbury.

Lipsett, S., & Marks, G. (2000). *It didn't happen here: Why socialism failed in the United States*. New York: W. W. Norton.

Lucas, J. (2015). Urban governance and the American political development approach. *Urban Affairs Review*. https://doi.org/10.1177/1078087415620054.

Maher, N. (2008). *Nature's new deal*. New York: Oxford University Press.

Mathews, J. T. (1997). Power shift. (changing role of central government). *Foreign Affairs, 76*(1), 50. https://doi.org/10.2307/20047909.

Mazower, M. (2012). *Governing the world: The history of an idea, 1815 to the present*. New York: The Penguin Press.

Merrifield, A. (2013). The urban question under planetary urbanization. *International Journal of Urban and Regional Research, 37*(3), 909–922. https://doi.org/10.1111/j.1468-2427.2012.01189.x.

Miller, D. L. (1989). *Lewis Mumford, a life*. New York: Weidenfeld & Nicolson.

Moisio, S. (2018). Urbanizing the nation-state? Notes on the geopolitical growth of cities and city-regions. *Urban Geography*, 1–4. https://doi.org/10.1080/02723638.2018.1454685.

Monbiot, G. (2000). *Captive state: The corporate takeover of Britain*. London: Macmillan.

Morgan, K. (2007). The polycentric state: New spaces of empowerment and engagement? *Regional Studies, 41*(9), 1237–1251. https://doi.org/10.1080/00343400701543363.

Newman, E. (2009). Failed states and international order: Constructing a post-Westphalian world. *Contemporary Security Policy, 30*(3), 421–443. https://doi.org/10.1080/13523260903326479.

Noortmann, M., Reinisch, A., & Ryngaert, C. (2015). *Non-state actors in international law*. Oxford; Portland, Oregon: Hart Publishing.

Orren, K., & Skowronek, S. (1996). Institutions and intercurrence: Theory building in the fullness of time. *Nomos XXXVII, Political Order, 38,* 111–146.

Orren, K., & Skowronek, S. (2004). *The search for American political development*. Cambridge: Cambridge University Press.

Peck, J. (2008). Remaking laissez-faire. *Progress in Human Geography, 32*(1), 3–43. https://doi.org/10.1177/0309132507084816.

Pred, A. (1977). The choreography of existence: Comments on Hagerstrand's time-geography and its usefulness. *Economic Geography, 53,* 207–221.

Robinson, J. (2006). *Ordinary cities: Between modernity and development*. London and New York: Routledge.

Ruffin, F. (2013). Municipal international relations: The South African case of metropolitan eThekwini. *Loyola Journal of Social Sciences, XXVII*(1), 119–141.

Sarkar, J. (2015). Debating a post-Westphalian international order. *Mainstream Weekly, LIII*(15).

Sassen, S. (1991). *The global city: New York, London, Tokyo*. Princeton: Princeton University Press.

Schroeder, H., & Bulkeley, H. (2009). Global cities and the governance of climate change: What is the role of law in cities. *Fordham Urban Law Journal, 36,* 314–360.

Scott, A. (2001). Globalization and the rise of city regions. *European Planning Studies, 9*(7), 813–826.

Seabrooke, L. (2002). *Bringing legitimacy back in to neo-Weberian state theory and international relations* (Working Paper No. 6). Department of International Relations. Canberra: Australian National University.

Sellers, J. M. (2002). *Governing from below: Urban regions and the global economy*. Cambridge and New York: Cambridge University Press.

Skowronek, S. (1982). *Building a new American state: The expansion of national administrative capacities, 1877–1920*. Cambridge [Cambridgeshire] and New York: Cambridge University Press.

Slaughter, A.-M. (1997). The real new world order. *Foreign Affairs, Sept/Oct,* 183–197.

Sluga, G., & Clavin, P. (2017a). Rethinking the history of internationalism. In G. Sluga & P. Clavin (Eds.), *Internationalisms: A twentieth century history* (pp. 3–16). Cambridge: Cambridge University Press.

Sluga, G., & Clavin, P. (Eds.). (2017b). *Internationalisms: A twentieth-century history*. Cambridge, UK: Cambridge University Press.

Sowell, T. (1998). *Conquests and cultures: An international history* (1st ed.). New York: Basic Books.

Steffek, J. (2015). Fascist internationalism. *Millennium, 44*(1), 3–22. https://doi.org/10.1177/0305829815581870.

Steger, M. (2002). *Globalism*. Landham, MD: Rowman and Littlefield.

Stone, C., & Whelan, R. (2009). Through the glass darkly: The once and future study of urban politics. In R. Dilworth (Ed.), *The city in American political development*. New York: Routledge.

Swenson, P. (2002). *Capitalists against markets: The making of labor markets and welfare states in the United States and Sweden*. New York: Oxford University Press.

Taylor, P. J. (2005). New political geographies: Global civil society and global governance through world city networks. *Political Geography, 24*(6), 703–730.

Thrift, N., & Amin, A. (2002). *Cities: Reimagining the urban*. Cambridge: Polity.

Toly, N. (2017). The new urban agenda and the limits of cities. *The Hedgehog Review, 19*(2), 36–44.

Walker, R. (2015). Building a better theory of the urban: A response to 'Towards a new epistemology of the urban?' *City, 19*(2–3), 183–191. https://doi.org/10.1080/13604813.2015.1024073.

Waterman, P., & Wills, J. (2001). *Place, space and the new labour internationalisms*. Oxford, UK; Malden, MA, USA: Blackwell.

Waterman, P., & Wills, J. (2012). Place, space and the new labor internationalisms: Beyond the fragments? In P. Waterman & J. Wills (Eds.), *Place, space and the new labour internationalisms*. London: Blackwell.

CHAPTER 3

The Greening of Internationalism: From Growing Impact Crisis to Stagnated Reconciliation Project

Abstract How and in what ways did we start to see the environment as a global political issue? Is it plausible to argue, with Liam Leonard (Global ecological politics, Emerald Group, Bingley, 2010, p. 1), that environmentalism is "the most significant contemporary global movement" to have emerged in recent decades? This chapter explores selective moments in the rise of an eventual stagnation associated with the greening of internationalism. Major concerns include the growing recognition of the environment in world political discourses and especially deepening concerns with threats posed by irreversible ecological limits across various global "commons." This chapter outlines key political efforts to address globally shared environmental problems over several decades of multilateral diplomacy. In addition, the discussion traces the recent stagnation and even reversal of the traditional Westphalian order in effectively addressing multiple green challenges to society and economy.

Keywords Ecological crisis · Global capitalism · Green internationalism Stockholm conference · Brundtland report · Agenda 21 Sustainable development goals

Y. Dierwechter, *The Urbanization of Green Internationalism*, Cities and the Global Politics of the Environment, https://doi.org/10.1007/978-3-030-01015-7_3

The environmental movement is perhaps the most significant contemporary global movement to have emerged in recent decades.

—Liam Leonard (2010, p. 1)

Sustaining...Human Development

Sometime after 1500 or so, the planet's diverse "economy-worlds"—Chinese (Meso-)American, African, European, Indian, Islamic, etc.—were steadily integrated, unevenly and often violently, into an incipient but irreversible "world-economy" (Bairoch, 1988). Through mercantile capitalism and transoceanic colonialism, and specifically through the widening of circulation—or what Fernand Braudel (1979) famously titled "the wheels of commerce"—the planet's terrestrial diversity was integrated and transformed by the territorialities of production and consumption. Spices, slaves, metals, foodstuffs, animals, diseases—all these moved through new routes and flows, creating novel cultural formations, technological diffusions, and economic/ecological articulations. Sometime after roughly 1800, the industrialized carbonization of these economic and ecological flows hyper-accelerated what many now see as the Anthropocene, a distinctive period in the earth's history defined mainly by human modifications of various ecosystems, especially the global climate.

Internationalism addresses many cross-border challenges, e.g., nuclear proliferation, failed states, terrorism, human rights, labor practices, legal norms, technical protocols, involuntary immigration, drug and sex trafficking. But the interlocking crises plaguing the long-term viability of the planet's biophysical commons must surely top this depressing list of large-scale human problems. Human development is fundamentally about the economic transformation of nature into societies—the ongoing construction and extension of socio-natural "hybrids" (Smith, 2003). But can human development in the urbanized Anthropocene be *sustained* (Harari, 2015)? Can humans build just societies and more responsible economies without obliterating nature's core resiliencies? For Foster, Clark, and York (2010), writing in the radical tradition of eco-Marxism, the answer is a contingent no: globalized capitalism is systematically destroying the Earth's protective biosphere, creating an ecological "rift" between humans and nature through unrelenting, self-reinforcing commodification.

Evidence of that rift includes, for them and others, global warming, species extinction, pollinator decline, nonrenewable energy usage, habitat destruction, invasive species, air quality, genetically modified food, overgrazing, monoculture, meat production, land pollution, desertification, soil erosion, soil salinization, urban sprawl, habitat fragmentation, high-level radioactive waste management, environmental impact of the coal industry, acid rain, ocean acidification, mercury in fish, smog, overconsumption, overfishing, deforestation, acid mine drainage, hydraulic fracturing, mountaintop removal mining, herbicides, pesticides and toxic waste—to name just a few that come to mind. So put more simply: the search for global sustainability may be the most important journey human beings ever take. Neil Brenner and Christian Schmitt's thesis of planetary urbanization, discussed in Chapter 1, suggests that rather little of this journey lies outside the ambit of urban studies, particularly as urbanism merges with international relations. Hence classic fields within urban studies—including planning and architecture—now ask how ecological design can help to ameliorate global ecological decay.

Such ideas, though, are recurrently dismissed (Caprotti et al., 2017). For some, it is too late for sustainable development (Meadows, 2008). It has stagnated before it has gathered sufficient momentum. Intergenerational equity, for instance, has been a major feature of sustainability discourses and environmental ethics since the publication of the Brundtland Commission's report, *Our Common Future*, in the mid-1980s (Kanninen, 2012). But it has proven exceptionally hard to think of ourselves as somebody else's ancestors, as faded photographs in the future scrapbooks of distant strangers—even within our own bloodlines. We are alive *today*. We want to enjoy what little time we do have. We see tensions between "gratification now and justice later" (Holden & Scerri, 2013, p. 445). This micro-sociological condition paradoxically generates circuits of ecological overconsumption and theaters of social inequality.

Is it too late? How *did* we get here? How and in what ways did we start to see the environment as a global political issue? Is it truly plausible to argue, with Liam Leonard (2010, p. 1), that environmentalism is "the most significant contemporary global movement" to have emerged in recent decades? With these questions in mind, this chapter traces the rise of, and recent stagnation associated with, what I am calling the *greening* of the "internationalism" just discussed in Chapter 2. Major concerns include the growing recognition of the environment in world political discourses and deepening concerns with threats posed by irreversible ecological limits across various global "commons." This chapter outlines political efforts to address globally shared environmental problems over several decades of multilateral diplomacy (e.g., Broadhead, 2002). It then traces the stagnation and reversal of the traditional Westphalian order in addressing green challenges to society and economy (Amen, Toly, McCarney, & Segbers, 2012). Ultimately, the purpose in what follows is to help establish the basis for some of the many novel and highly variegated urbanization(s) of green internationalism mapped out in greater detail in Chapter 4.

The Greening of Internationalism

Environmentalism is not easy to define and is appropriately subject to considerable debate. It has relatively long discursive bloodlines over time—longer than most people think. Seen as the intellectual and political awareness of ecologies of interconnection between "social" and "natural" worlds, environmentalism links us backward in time to iconic experiences of place and famous personages or groups worried about environmental futures (which then soon enough became our various "pasts"), e.g., the biologist-planner Patrick Geddes in Edinburgh; John Muir in Hetch Hetchy, California; Henry David Thoreau on Walden Pond; the forest-protecting Bishnois Hindus of Khejarli, Rajasthan, in the early eighteenth century. Many others could make this list.

Lumley and Armstrong (2004) trace the specific concept of sustainability to the early nineteenth century. But in so far as sustainability incorporates concerns with equity and wider problems of social justice, environmentalism is arguably an ancient ideal—present in Hellenic, Roman, Chinese, and Islamic intellectual life. No less than Montesquieu, reflecting on classical philosophy, demography and ancient history's relationship to an emerging European modernity, inaccurately but revealingly wondered:

> How can the world be so sparsely populated in comparison with what it once was? How can nature have lost that prodigious fertility of primitive times? Could she already be in her old age, and will she fall into her dotage? (cited in Glacken, 1973, p. 134)

Environmental issues have constituted the basis of cross-border if not truly globalized treaties, protocols, and amendments since at least the mid-nineteenth century, particularly in Western Europe. Diplomats and other public officials made pioneering efforts in the 1850s, for instance, to regulate the flow of water from Lake Constance, Switzerland, and during the 1860s to begin to monitor more assiduously various pollutants along the Rhine. Many other international treaties and protocols soon dealt with protecting fisheries, seals, and plants in shared oceanic, lake, and riverine waters. After World War I in a diplomatic process that is still unfolding three nations—Estonia, Sweden, and Czechoslovakia—agreed to regulate white lead. (By 2017, more than sixty countries around the world had officially joined the *Convention Concerning the Use of White Lead in Painting* first formulated in 1921.)

Such cooperation reflected a myriad of intellectual, cultural, and social changes—in science, poetry, philosophy, and politics—that steadily elevated the fragility of natural environments as a human challenge and in time an abiding global concern that is today bordering, for scientists in particular, on existential crisis (Kolbert, 2014). Environmental thinkers and actors, from Henry David Thoreau to George Perkins Marsh, changed wider intellectual landscapes and ways of thinking. The closing of the American frontier in the late nineteenth century, captured by Frederick Jackson Turner's "Frontier Thesis" on the geographical sources of American democracy, prompted concerns with environmental limits and helped to ignite a famous debate about conservation versus preservation between Gifford Pinchot and Join Muir, respectively (Mayer, 1997). During Franking Roosevelt's New Deal, new kinds of governance spaces emerged that began to link the environment with the economy and even social equity concerns—all early antecedents of sustainability thinking (Dierwechter, 2010). Recent research additionally documents the neglected histories, sociologies, and geographies of African–American environmentalism and black environmental thought, to take one example. This work offers correctives to a skewed, racialized historiography of green action (Chesney, 2008).

A raft of public-intellectual and legal-institutional interventions and changes after World War II—and especially during the 1960s—left deep marks on the widening landscapes of globalizing environmental discourses (Caradonna, 2014). Ironically, as Caradonna (2014) shows in her recent history of sustainability, many came out of the USA, where more recent commitments to global ecological politics have wavered between (insufficient) progress under Clinton and Obama to (profound) regression under Bush and especially Trump. Classic concerns with war, security, and the legal-institutional search for peace and prosperity between Great Powers—the hard core of world politics—dominated international action throughout the post-war era. But even as the Cold War between the USA and the Soviet Union, which peaked with the Cuban Missile Crisis in October 1962, shaped cartographic imaginations and real practices, a parallel set of bilateral and multilateral agreements pushed global environmental problems onto the world stage.

Admittedly, some of these new practices reflected Cold War geopolitics. We associate the politics of the atmosphere today with greenhouse gas emissions and the policy urgency of multi-scalar carbon reduction. In 1963, however, the lingering anxieties of the Cuban debacle—and the near Armageddon of large-scale radioactive fallout—helped countries to ban nuclear testing in the atmosphere, in outer space, and underwater (though not yet underground) (Seabrook, 1981). New international diplomatic activity around several ecological spaces—oceans, lakes, hazards, air, etc.—nonetheless slowly shifted, year after year, the overall terrain of world politics as the post-World War II era played out.

As Caradonna (2014) further observes, the son of the University of Wisconsin ecologist, Aldo Leopold, posthumously published his father's influential *Sand County Almanac* in 1949, which argued for a new "land ethic" that challenged dominant land-use narratives and property norms. Within American urban studies, she reminds us, Jane Jacobs's devastating critique of modernist planning, *The Death and Life of Great American Cities*, published in 1961, elevated popular appreciation of architectural diversity, mixed-use development, and pedestrian-oriented public spaces that anticipated key themes now more common in sustainable urban development literatures (Trudeau, 2013). A year later, Rachel Carson's *Silent Spring* drew attention to the environmental effects of pesticides. In 1968, Garrett Harden's article, "Tragedy of the Commons," punctured a major hole in the presumed public benefits of private self-interest—a linchpin postulate of unfettered capitalism since Adam Smith's writings in the late eighteenth century. Renewing themes in the neo-Malthusian school, Paul Erlich, also in 1968, warned of an approaching "population bomb." Barry Commoner's book, *The Closing Circle*, asserted in 1971 that capitalist technology—not rising population—was doing the most to degrade global ecosystems. National-level environmental legislation, such as the two Clean Air Acts of 1956 and 1965 in the UK and USA, respectively, gathered considerable momentum. In 1972, Richard Nixon approved the use of Agent Orange in the jungles of Vietnam; yet he created the Environmental Protection Agency and signed the National Environmental Policy Act at home (Caradonna, 2014).

Like most developments of consequence, then, tracing the intellectual and practical emergence of global concerns with sustainability is like pealing a large and slippery onion. Each layer reveals yet another layer. Most scholars nonetheless emphasize the intensive greening of internationalism, as I am calling it here, to the early 1970s.

From I = PAT to the 3Es: Embracing Agenda 21

A key initiating event—indeed a "landmark" in this particular history (Broadhead, 2002)—appears to be the United Nations Conference on the Human Environment, held in Stockholm in mid-1972 (Eardley-Pryor, 2014). Although the conference helped to create the United Nations Environment Program (UNEP), based in Nairobi, multilateral discussions revealed pervasive tensions between state-centric/progressive and more radical/society-based philosophies and concerns. These reflected the different traditions of internationalism discussed in Chapter 2.

As the UN delegates gathered inside, NGO critics outside the conference—notably the Dai Dong Group—questioned the efficacy of international systems of decision-making and modes of regulation (Broadhead, 2002). No doubt the Dai Dong Group's more radical critique was influenced by the green intellectual ferment just recounted, and perhaps by the contemporaneous publication of the now famous "Club of Rome" report, *The Limits to Growth.* This report effectively modeled large-scale ecological collapse on a planetary scale sometime before the year 2100 absent major and quite widespread changes in economic, political, and social life.

That said, the Stockholm conference is important because, as Caradonna (2014, p. 142) specifically notes, "the emergence of sustainability within the milieu of the UN signaled the increasingly international character of the movement." Unfortunately by the 1980s, the USA under Ronald Reagan had retreated from environmental leadership, both nationally and internationally, while the UN system built precariously on its modest green foundation from the 1970s (Mazower, 2012). In parallel, many of the really interesting experiments in sustainability policies coalesced in countries like Germany, the Netherlands, and Sweden (Beatley, 1999).

Such experiments and policies arguably reflected the internal and external impacts of theoretical and positive notions of the "green" or "eco" state as a legitimately new form of "stateness." For Eckersley (2004, p. 4), a distinctly green democratic state, at least in theory, provides "an alternative to the classical liberal state, the indiscriminate growth dependent welfare state, and the increasingly ascendant neoliberal competition state" (Fig. 3.1). Yet as first argued in Chapter 2, states are not singular things; they are complex assemblages of sometimes complementary and often competing institutions, discourses, and rationalities of territorial governance and social control.

Accordingly, the green democratic state currently contends with the neoliberal competition state—and these tensions *seep into* the interstate system. The state's "classic" functions include—and still include—distribution, arbitration, and security services as manifested empirically through modes of enforcement (taxes, penalties, legal norms, incentives, etc.) and reproduced through processes of legitimation (political rituals, ideologies, cultural codes, etc.). The corporate capture the state discussed in Chapter 2 has reshaped key state functions into neoliberal

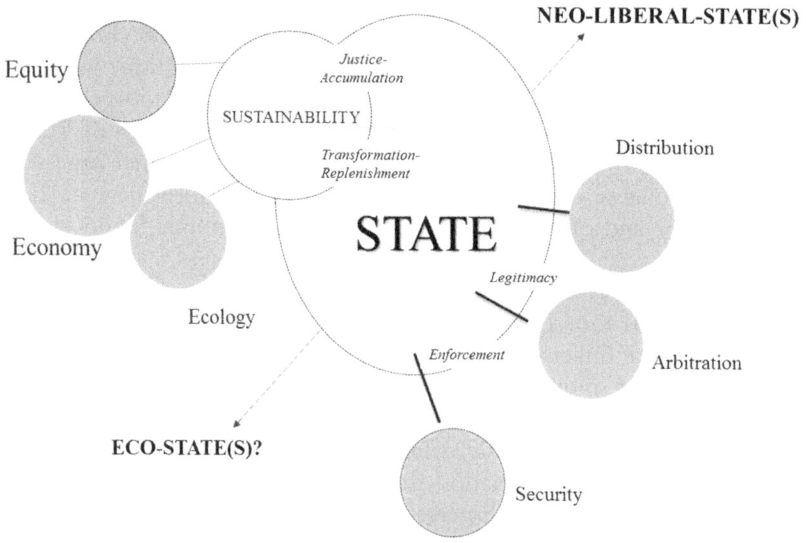

Fig. 3.1 State functions and contending state forms

forms through, inter alia, upward income distribution from centralized accumulation practices. Yet eco-state functions and even the global emergence of select eco-states (Duit, 2014) reflect the impacts of how environmental thought has widened into concerns with development.

In fact, major actors in the Global South like Brazil, India, and eventually China, helped to reshape the inherited *environmentalism* of sustainability into a much broader "development theory" for how, in general if contradictory terms, the ongoing transformation of nature's systems into sustainable economies and just societies might proceed in ways that also reduced global-scale inequalities inherited from colonial relationships and forms of dependency (Preston, 1996). Intellectually, the 1980s saw the conceptual emergence of the normative and analytical heuristic that "development" at all scales should (and could) be about the *concomitant pursuit* of ecological resiliency social equity, and economic growth. The new language of sustainability thus revealed its own set of tensions between justice and accumulation, for instance, and specifically between the transformation of nature into economies while still being concerned with replenishment.

Soon simplified and stripped down as it circulated through global policy fora, the so-called 3Es model (economy, equity, ecology) displaced the influence of the quintessential formula of the 1970s, codeveloped by Paul Erhlich, Barry Commoner, and John Holdren as I=PAT. The I=PAT formula sought to calculate the impact of human activities on the environment as the multiplication of population, affluence, and technology (Fig. 3.2).

The intellectual shift from the presumed alarmism of I=PAT to the preferred geopolitics of a more pliable 3Es is captured by the second key moment in the environmentalization of internationalism: the Brundtland Report in 1987, which was published officially as *Our Common Future*. Written under the auspices of the UN's World Commission on Environment and Development, the Brundtland Report attempted to reconcile contending intellectual streams and somehow ameliorate global tensions exposed at Stockholm and later UN conferences. Bruntland's chief innovation, then, was to formalize and diffuse the still-new concept of sustainable development in such a way that "the opposing concerns of Northern supported environmental conservation and Southern desired economic development" could productively gather (Harlow, Golub, & Allenby, 2013). The "impact crisis" that characterized environmental discourses in the 1960s and early 1970s, once internationalized after

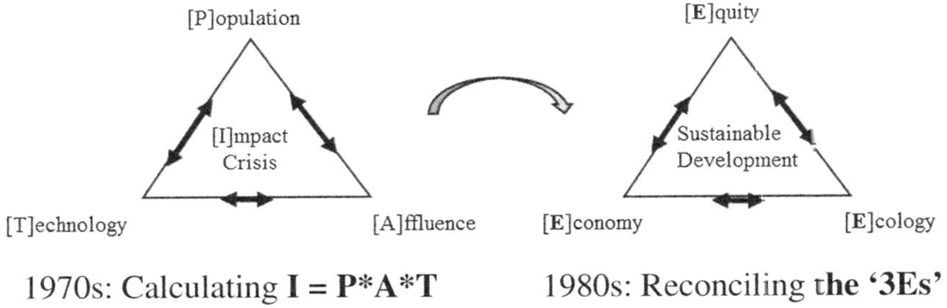

Fig. 3.2 Discursive shift in environmental thought from early 1970s to late 1980s

1972, morphed into a "reconciliation" problem by the late 1980s. So imagined, sustainable development actually occurred—or "arrived" in the world—when political communities figured out how specific projects or programs concomitantly generated measurable improvements in all three domains of socio-natural development.

Hypothesized instances of the 3Es in applied sustainability practice are now legion in both the policy and scholarly literatures and at all scales of governance and development. A few cases help to illustrate this point. "Project-level" efforts to reconcile the three sustainability goals include initiatives to provide affordable housing developments with green building materials, as in Boston, where green affordable housing policy is considered potentially more cost-effective than conventional affordable housing on a long-run, life-cycle basis "because of the lower utility and replacement costs" (Burke, Nelson, & Rickerson, 2007, p. 3). Other examples include the American civil engineering profession's engagement with sustainable street design (Greenberg, 2009). At the level of wider programs and territorial policy frameworks, a prominent example of the 3Es philosophy of sustainable development is found in California's statewide workforce development strategies in the late 1990s and early 2000s. According to one White Paper, the various influences of California's workforce development on the 3Es "are fluid, inter-connected, and often evident in almost every state sponsored workforce development effort" (Feinberg, 2003, p. 6). In Japan, as a final case, efforts to reconcile the 3Es are seen in recent reforms to national energy plans (Duffield & Woodall, 2011).

Nonetheless, problems with the 3Es reconciliation thesis are well-known. From a liberal-progressive perspective, as many scholars have argued, the third E in the 3Es model—equity—is typically ignored, diminished, or weakly addressed (Clark, 2015). Equity is often delivered with vague reference to co-benefits associated with "community," and less frequently with the redistributive politics of social democratic values and priorities. Much of this literature suggests the key problem is simply to push forward an equity agenda (Vojnovic, 2013). As a normative theory of societal development, the triangle must be more or less equilateral in all three corners rather than skewed overwhelmingly toward, say, economic growth. The more equilateral, the more sustainable. Thus, 3E sustainability also requires a vibrant democratic polity capable of negotiating the particulars of reconciliation within any given cultural setting. In more radical accounts, in contrast, post-Brundtland initiatives in global sustainably do not represent a green turn in international politics. They represent an about-face from the far-reaching demands and implications of the late 1960s and early seventies—arguably a more tumultuous period of world intellectual and political history overall.

The fast-approaching "impact crisis" signaled by Erlich, Commoner, the Club of Rome, and the Dai Dong Group, among many others, was in the end too easily neutralized by the "reconciliation" illusions of sustainable development. Tulloch and Neilson (2014) suggest that the earlier and more radical discourses associated with the impact crisis—different but all emphasizing a stark ethic of "survivalism"—were incorporated and subordinated to the neoliberal hegemony of the 1980s, when the slower-growing economies of the industrialized West generated a renewed politics of deregulation and privatization. What happened, in essence, is that global sustainably now meant sustaining (rather than transforming) capitalism's "expansionist logic" (p. 27). Like vague notions of community, the global reconciliation project of sustainable development seemed more appealing and, on the surface, discursively unassailable. Whereas the I = PAT diagram suggested a degenerative triangle that produced *peril*, the Brundtland discourse counter-suggested a happy triangle that produced *progress*. The Brundtland project of the 3Es depended not only on making the economy and ecology "equivalential," but on implementing sustainability through (and for) fully neoliberalized economies (ibid.).

Now well into the twenty-first century, it may not be that long—perhaps a few decades from now, perhaps much sooner—before we really began to see who is fundamentally right: the impact radicals or the reconciliation reformists. For the moment, reformists working within the post-Brundtland parameters of a more environmentally responsible vision of globalized capitalism dominate the sustainability narrative at all scales, including the international scale.

The practical mechanics of the 3Es are supported intellectually by a range of theoretical assurances, especially those associated with forms of ecological modernization and attendant concepts like industrial ecology or ecological design—and more recently, "smart city" experiments (Evans & Karvonen, 2014). Developed originally in the 1980s by German scholars, ecological modernization, for instance, refers in general terms to how environmental problems "come to be framed as issues that are politically, economically and technologically solvable within the context of existing institutions and power structures and continued economic growth" (Bailey, Gouldson, & Newell, 2011, p. 683). As two of its early adherents admit, ecological modernization does not aim for a fundamentally different organization of capitalist society, but for modernization "with an ecological twist" (Mol & Janicke, 2009)—or what Huber called "superindustrialisation," wherein "...the dirty and ugly industrial caterpillar will transform into a[n] ecological butterfly" through the adoption of improved technology (quoted in Murphy and Gouldson, 2000, p. 34).

The twist of ecological modernization and similar theories is not always focused sufficiency on social equity concerns; but it expeditiously offered the international community a legal-political way out of what seemed the core tautology of the sustainable development project, namely, that "environmental protection must not impede development, which, in order to be sustainable, should necessarily take account of environmental concerns" (Momtaz, 1996, p. 262). It offered a strategic way to rethink how "existing institutions and power structures" (op. cit.), especially international institutions and structures, could now implement the values, plans, and goals first promulgated at Stockholm in 1972. That was obviously important because, as Momtaz argues, the agencies and commission of the United Nations over the 1970s and 1980s became the leading catalysts for global environmental protection.

Exactly twenty years after Stockholm—in June 1992—the largest gathering yet of nation-states ostensibly committed to improving global environmental stewardship through sustainable development programs met in Rio de Janeiro. In key respects, Rio represents today the zenith of a certain kind of green internationalism; a high-water mark of the Westphalian system's state-to-state environmental diplomacy; a particular kind of environmentalized international politics. For pessimists, it was the beginning of the end, a final shuttering of the open doors of the 1970s. There is no butterfly, in this reading, only the caterpillar. For optimists, in contrast, Rio was more the end of the beginning—a critical inflection point to consolidate better the world community's limited but tangible environmental accomplishments in the antecedent decades. The problem now was to push harder for implementation as the twenty-first century approached.

The United Nations Environment Program—the main institutional creation of Stockholm—now provided a kind of scientific internationalism, to again use Mark Mazower's language. UNEP applied its considerable expertise to monitoring global problems like atmospheric pollution, deforestation, desertification, the thinning ozone layer, and so on. Armed with new data, it also placed more pressure on member states in regard to strengthening their various national environmental laws and other reforms (Momtaz, 1996). Rio rode this tradition. UNEP's scientific remit to global governance sharpened administratively with a new Commission for Sustainable Development, which was one of the key institutional mechanisms to pursue what appears today as one of the three key outcomes of Rio: Agenda 21.

Agenda 21 was a complex document published by the United Nations within a year of the Rio Conference. It provided an analytical context, rationale, and most importantly, large set of policy actions for signature states to undertake, albeit voluntarily, in their respective national settings (United Nations Conference on Environment and Development, 1993). After a short preamble that emphasizes the stakes involved, section "Sustaining…Human Development" outlines socioeconomic relationships between sustainability and extant patterns in global poverty, demography, human settlements, health, and "decision-making." Section "The Greening of Internationalism" elaborates upon key challenges associated with "the conservation and management of resources for development"—some natural (e.g., oceans, forests, lakes, mountains, atmosphere); others, socio-natural (e.g., biotechnology, waste products). Interestingly, section "From I=PAT to the 3Es: Embracing Agenda 21" turns to the theme of participation and capacity building, highlighting the "roles" of women, children, indigenous peoples, non-governmental organizations, local authorities, workers and trade unions, businesses and industries, scientists and the technological community, and farmers. Section "From I=PAT to the 3Es: Embracing Agenda 21" concludes the sprawling document with a multidimensional presentation of the "means of implementation" that, the authors suggest, is necessary to redress the global problems members collectively face. Figure 3.3 visualizes these means as an "octagon" of intersecting challenges.

The scale of the problems, the profound stakes at play, and the complexity of post-agreement implementation have not necessarily generated as much serious assessment as might be expected, particularly from fields like international relations and law. Most research on Agenda 21 is from environmental science disciplines, public administration, urban studies, and geography. According to the Web of Science, most of this research has focused on Local Agenda 21 and especially the role of

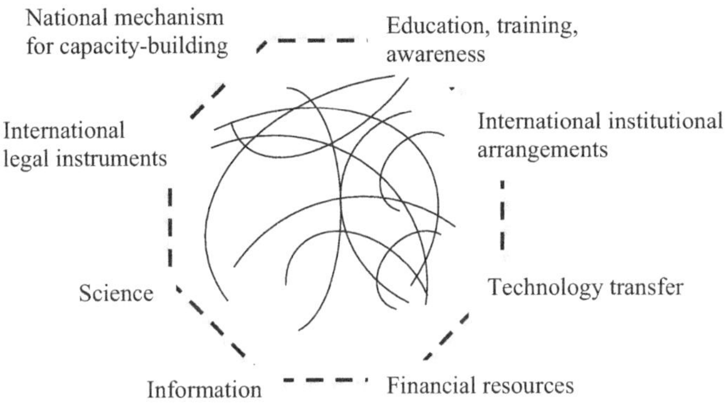

Fig. 3.3 Agenda 21: means of implementation (*Source* UNCED, 1993)

local authorities rather than labor unions, women, corporations, and so on. More will be said of this particular focus in Chapter 4. For the moment, it is sufficient to note that scholarly efforts to understand the linkages and empirical impacts of, for example, global developments in international legal instruments on global patterns in technology transfer are considerably thinner than required. Critiques and concerns with the overall design and conception of Agenda 21 as a global program for sustainable development are more common (Corbiere-Nicollier, Ferrari, Jemelin, & Jolliet, 2003), including concerns that focused on particular areas of policy in the original framework such as: poverty alleviation (Heyd, 1995); education (Real, 1999); NGO activism (Alyanak & Cruz, 1997); or the utilization of science in sustainability programs (Funtowicz, Ravetz, & O'Connor, 1998).

Early generalized assessments of Agenda 21's overall effects include Langeweg's (1998) suspicion that rather than shaping a "common future," as the Brundtland Commission had hoped, global efforts suggested instead a "common failure." "Although we are now 5 years along the road marked out by Agenda 21," he concluded,

> the *perspectives* for achieving the objectives stated are still slight. Despite economic growth, global inequity is on the increase. The conversion of natural areas to agricultural land will greatly increase due to the population growth, especially in Asia and Africa. In North America and Europe this trend seems reversed: life expectancy is longer and people's health has

> improved. The developing world is catching up. However, *there is more needed* to bring the world population's welfare and social security levels up to par. At the same time the environmental pressures on the world's natural areas will have to be brought down substantially. (p. 227, emphasis added)

The other principal achievements of the Rio Conference were two international environmental treaties, viz. (1) the Convention on Biodiversity and, given the dramatic rise of climate change concerns since the early 1990s, the arguably much more impactful and far-reaching (2) United Nations Framework Convention on Climate Change.

Both treaties are landmark legal documents. They have addressed questions of how to reconcile economic development with the environment while also attending to post-Brundtland concerns with inter- and intra-generational equity. They represent the first legal documents in international history that sought to address "sustainable development." Along with Agenda 21, they signaled that sustainable development now constituted "the common goal of humankind" (Hodas, 2011, p. 260). The irony of this claim is that climate change has superseded other global environmental and social security problems—including biodiversity, soil loss, desertification, poverty and inequality and, even, to some extent, the organizing meta-narrative of sustainability and sustainable development. In so far as a "common goal" does exist, in other words, a good case can be made for the rise of climate change, both with respect to mitigation and adaptation. As one local official in Erlangen, Germany put it when asked by the author about "climate change" at the local scale versus planning for sustainable development: "Aren't they really the same thing?"

As late as the latter 1980s, major scholarly discussions of international political economy could (almost) entirely ignore the rise of environmental concerns in global affairs (Gilpin, 1987). The surprising denouement of the Cold War, the implications of globalization and not least the global politics that Rio set in motion arguably changed all this. Just as traditional states started to relocate and elevate "the environment" as part of global politics, global politics changed too. The limits of the liberal Westphalian world order increasingly drew serious attention. So too did the hypothesized rise of cities as global economic and political actors.

Conclusions

The intellectual bloodlines of environmentalism and the ethical concept of sustainability can be traced back centuries, but the greening of internationalism—the emergence of a distinctive global ecological politics—dates mainly to the late 1960s and 1970s. As individual states increasingly took on environmental(ist) concerns, as "eco-state" functions unevenly emerged, the interstate system in global affairs similarly turned its institutional diplomatic, scientific, moral, and legal attention to problems like acid rain, fisheries, species loss, desertification, and soil degradation.

By 1970, an expansive green literature on what appeared to some as an "impact crisis" had fully developed. Yet key moments in the history of the United Nations—Stockholm, Brundtland, and Rio—soon signaled a steady shift away from ecological discourses of impact to developmental promises of reconciliation. In particular, ecological resiliency should be—and could be—***reconciled*** with economic development in ways that concomitantly expanded social equity, a normative innovation for practical guidance that quickly circulated through the policy networks of international institutions, including the United Nations.

The reconciliation project of the 1990s, captured by the post-Rio Agenda 21 program, benefitted from the macro-historical backdrop of global optimism that attended the surprisingly peaceful denouement of the Cold War; the hypothesized march of democratic values; and, just as the Internet opened up new spaces of collaboration and communications, the sense that the Frankenstein technologies of the I=P*A*T formula could somehow mutate and eventually renew societies into something smarter, cleaner, and greener—that the ugly caterpillars of industrial capitalism could turn into beautiful butterflies within new economic ecosystems, magically modernized and institutionally updated for a twenty-first-century world shaped more by digits and apps than oil and coal. A new world could be re-spatialized by the 3Es.

That the impacts of population, affluence, and technology were still outpacing, outstripping, even now overwhelming the multi-scalar reconciliation of equity with economic growth and ecological resiliency were plainly evident long before the dark turn in global affairs associated with 9/11 and its geopolitical aftermath. As the iconic Norwegian leader, Gro Harlan Brundtland, warned the world community in an early 1990s era speech:

> We all recognize the signs of the global crisis now approaching. Global warming, depletion of the ozone layer, continued population growth, massive loss of species and biological diversity, acceleration of deforestation and desertification — these are all threats which will soon lead to breakdowns in vital support systems for life on earth. (Brundtland, 1991, p. 39)

Whatever global hopes that nascent notions of *ecological* security for human populations may have had to accelerate progress were quickly crowded out by the War on Terrorism and the American rejection of the Kyoto Treaty, if single events can be cited. But as in previous epochs, the closing of one set of dynamics opened up interests in other possibilities. Stagnation was unacceptable. The *urbanization* of green internationalism—partly neoliberal; partly eco-statist; partly both at the same time; definitely ascendant—now seemed to offer a new direction for global environmental politics. And so, a renewed love affair with cities as familiar but long neglected political spaces budded again as the late twentieth century crossed into the opening decades of the twenty-first century. As it did, I shall now argue, new urbanized geographies of a greener but stalled-out internationalism also emerged.

References

Alyanak, L., & Cruz, A. (1997). Implementing Agenda 21: NGO experiences from around the world. *Environment and Urbanization, 9*(2), 322–323.

Amen, M., Toly, N., McCarney, P., & Segbers, K. (Eds.). (2012). *Cities and global governance: New sites for international relations.* Farnham, UK: Ashgate.

Bailey, I., Gouldson, A., & Newell, P. (2011). Ecological modernisation and the governance of carbon: A critical analysis. *Antipode, 43*(3), 682–703.

Bairoch, P. (1988). *Cities and economic development: From the dawn of history to the present.* Chicago: University of Chicago Press.

Beatley, T. (1999). *Green urbanism: Learning from European cities.* Washington, DC: Island Press.

Braudel, F. (1979). *The wheels of commerce: Civilization and capitalism 15th–18th century.* New York: Harper & Row.

Broadhead, L.-A. (2002). *International environmental politics: The limits to green diplomacy.* Boulder, CO: Lynne Rienner.

Brundtland, G. (1991, February 14). *Environmental challenges of the 1990s: Our responsibility toward future generations.* The Tanner Lectures on Human

Values. Retrieved from https://tannerlectures.utah.edu/_documents/a-to-z/b/brundtland92.pdf.

Burke, P., Nelson, G., & Rickerson, W. (2007). Boston's green affordable housing program: Challenges and opportunities. *New York University Journal of Legislation and Public Policy, 11*(1), 1–32.

Caprotti, F., Cowley, R., Datta, A., Broto, V. C., Gao, E., Georgeson, L., Herrick, C., Odendaal, N., Joss, S. (2017). The new urban agenda: Key opportunities and challenges for policy and practice. *Urban Research and Practice, 10*(3), 367–378.

Caradonna, J. L. (2014). *Sustainability: A history*. New York: Oxford University Press.

Chesney, C. (2008). *African-American environmentalism: Issues and trends for teaching, research and extension*. Paper presented at the Black Environmental Thought: Land, Power and Sustainability Conference Kellogg Conference Center, Tuskegee University, Tuskegee, Alabama. Retrieved from http://e-research.tnstate.edu/extensionw/1.

Clark, T. P. (2015). Sustainability for whom? An examination of the equity pillar of sustainability at the local policymaking level. *Environmental Justice, 8*(2), 57–63. https://doi.org/10.1089/env.2015.0002.

Corbiere-Nicollier, T., Ferrari, Y., Jemelin, C., & Jolliet, O. (2003). Assessing sustainability: An assessment framework to evaluate Agenda 21 actions at the local level. *International Journal of Sustainable Development and World Ecology, 10*(3), 225–237.

Dierwechter, Y. (2010). Metropolitan geographies of US climate action: Cities, suburbs and the local divide in global responsibilities. *Journal of Environmental Policy & Planning, 12*(1), 59–82.

Duffield, J. S., & Woodall, B. (2011). Japan's new basic energy plan. *Energy Policy, 39*(6), 3741–3749. https://doi.org/10.1016/j.enpol.2011.04.002.

Duit, A. (Ed.). (2014). *State and environment: The comparative study of environmental governance*. Boston: MIT press.

Eardley-Pryor, R. (2014). *The global environmental moment: Sovereignty and American science on spaceship earth, 1945–1974* (Ph.D. dissertation). University of California, Santa Barbara. Retrieved from https://alexandria.ucsb.edu/downloads/n870zq96x.

Eckersley, R. (Ed.). (2004). *The green state*. Boston: MIT Press.

Evans, J., & Karvonen, A. (2014). 'Give me a laboratory and I will lower your carbon footprint!'—Urban laboratories and the governance of low-carbon futures. *International Journal of Urban and Regional Research, 38*(2), 413–430. https://doi.org/10.1111/1468-2427.12077.

Feinberg, T. (2003). *California's workforce development system*. Sacramento: State of California.

Foster, J., Clark, B., & York, R. (2010). *The ecological rift: Capitalism's war on the earth*. New York: Monthly Review Press.
Funtowicz, S. O., Ravetz, J. R., & O'Connor, M. (1998). Challenges in the utilisation of science for sustainable development. *International Journal of Sustainable Development, 1*(1), 2–10.
Gilpin, R. (1987). *The political economy of international relations*. Princeton, NJ: Princeton University Press.
Glacken, C. (1973). *Traces on the Rhodian shore: Nature and culture in Western thought from ancient times to the end of the eighteenth century*. Berkeley: University of California Press.
Greenberg, S. (2009). *Sustainable streets: Foundations for an emerging practice*. Paper presented at the Transportation Research Board 88th Annual Meeting, Washington, DC.
Harari, Y. N. (2015). *Sapiens: A brief history of humankind* (1st U.S. ed.). New York: Harper.
Harlow, J., Golub, A., & Allenby, B. (2013). A review of utopian themes in sustainable development discourse. *Sustainable Development, 21*(4), 270–280. https://doi.org/10.1002/sd.522.
Heyd, T. (1995). Agenda 21 provides a shaky foundation for fight against global poverty. *Alternatives: Perspectives on Society Technology and Environment, 21*(4), 33–33.
Hodas, D. (2011). International law and sustainable energy: A portrait in failure. In J. Benidickson, B. Boer, A. Benjamin, & K. Morrow (Eds.), *Enviornmental law and sustainability after Rio*. Cheltenhem: Edward Elgar.
Holden, M., & Scerri, A. (2013). More than this: Liveable Melbourne meets liveable Vancouver. *Cities, 31*, 444–453. https://doi.org/10.1016/j.cities.2012.07.013.
Kanninen, T. (Ed.). (2012). *Crisis of global sustainability*. London: Routledge.
Kolbert, E. (2014). *The sixth extinction: An unnatural history*. New York: Henry Holt.
Langeweg, F. (1998). The implementation of Agenda 21 'our common failure'? *Science of the Total Environment, 218*(2–3), 227–238. https://doi.org/10.1016/s0048-9697(98)00210-1.
Leonard, L. (2010). *Global ecological politics*. Bingley: Emerald Group.
Lumley, S., & Armstrong, P. (2004). Some of the nineteenth century origins of the sustainability concept. *Environment, Development and Sustainability, 6*, 367–378.
Mayer, J. (1997). Gifford Pinchot, John Muir, and the boundaries of politics in American thought. *Polity, 2*(2), 267–284.
Mazower, M. (2012). *Governing the world: The history of an idea, 1815 to the present*. New York: The Penguin Press.

Meadows, D. (2008). *30-year update of limits to growth finds global society in "overshoot," foresees social, economic, and environmental decline*. Retrieved from https://web.archive.org/web/20080517054744, http://www.clubofrome.org/docs/confs/meadows_abstract_21_08_04.pdf.

Mol, A., & Janicke, M. (2009). The origins and theoretical foundations of ecological modernisation theory. In A. Mol, D. Sonnefeld, & G. Spargaren (Eds.), *The ecological modernisation reader* (pp. 17–27). London: Routledge.

Momtaz, D. (1996). The United Nations and the protection of the environment: From Stockholm to Rio de Janeiro. *Political Geography, 15*(3), 261–271. http://dx.doi.org/10.1016/0962-6298(95)00109-3.

Murphy, J., & Gouldson, A. (2000). Environmental policy and industrial innovation: Integrating environment and economy through ecological modernisation. *Geoforum, 31*(1), 33–44. https://doi.org/10.1016/S0016-7185(99)00042-1

Preston, P. (1996). *Development theory: An introduction*. Oxford: Blackwell.

Real, W. (1999). Challenges of Agenda 21 to the environmental education. *Fresenius Environmental Bulletin, 8*(5–6), 230–235.

Smith, R. (2003). World city actor-networks. *Progress in Human Geography, 27*(1), 25–44.

Trudeau, D. (2013). New urbanism as sustainable development? *Geography Compass, 7*(6), 435–448.

Tulloch, L., & Neilson, D. (2014). The neoliberalisation of sustainability. *Citizenship, 13*(1), 26–38.

United Nations Conference on Environment and Development. (1993). *Agenda 21: Programme of action for sustainable development*. New York, NY: United Nations Department of Public Information.

Vojnovic, I. (2013). Advancing toward urban sustainability: The pursuit of equity. In I. Vojnovic (Ed.), *Urban sustainability: A global Context* (pp. 1–34). East Lansing: Michigan State University Press.

CHAPTER 4

Romancing the City: Three Urbanization(s) of Green Internationalism

Abstract This chapter builds on the conclusions of Chapter 3, using the metaphor of "romance" to assess the rise of cities as they have engaged states and international organizations, and vice versa. While international affairs have environmentalized, global environmentalism in turn has strongly ***urbanized*** in recent decades. The discussion thus presses the case for how urban space was steadily reconceptualized after the denouement of the Cold War as a "global solution" to ecological challenges. One major implication is that political ecologies have now "delocalized" and "upscaled," a process that has caused its own tensions and political contradictions. Attention is paid to signature initiatives like Local Agenda 21, Sustainable Development Goals (SDGs), the rise of inter-municipal policy networks, and recent "smartness" discourses. In particular, the chapter identifies three distinctive kinds of urbanizations: international, transnational, and smart. Each section considers major world cities to help illustrate synoptic themes. Cape Town, Los Angles, and Melbourne receive special treatment, respectively.

Keywords International urbanization · Transnational urbanization
Smart city urbanization · Cape Town · Los Angles · Melbourne

Y. Dierwechter, *The Urbanization of Green Internationalism*,
Cities and the Global Politics of the Environment,
https://doi.org/10.1007/978-3-030-01015-7_4

I could not seriously sit down to write a serious romance under any other motive than to save my life.

—Jane Austen (Le Faye, 1995)

Introduction

The rise of cities in the actual practices of global environmental politics is a romance story about the recent internationalization and complex rescaling of green urbanism. Some cities are putatively more romantic than others. Paris features, as always (Scruggs, 2017); but so too do a host of other localities (Girardet, 1999; Moore, 2007; Nwaka, 1996). The list is always growing, always amended, always contested as novel politics and fresh spaces of policy agency mature and transform. Some cities—some of our mental models of green urbanism—are by now entirely expected and frequently evoked: the superstars of Copenhagen, Barcelona, Curitiba, Vancouver, Singapore, or Melbourne, for instance; still other models are imagined through exotic, post-colonial discourses of glamor: Dakar, for example, or Quito. The romancing of these (and other cities) is part fantasy, it would seem, but also part reality—and even part rational expectation. As one project stagnates, another entices.

In itself, the post-Westphalian romancing of cities as presumed agents of global sustainability tells us only that the institutional arrangement for global Westphalian space—the world forged after 1815—has stagnated, as discussed in previous chapters. Beyond the colorful metaphor, however, lie multiple possibilities that are hardly predetermined. Romances are *real*, after all, but they can either become romanticized and even delusional as the initial shine wears off; more positively and productively, they can lead to long-lasting, deeper, more realistic, more consequential relationships that make a tangible difference for those involved.

This chapter considers the remarkable and rather recent urbanization of green internationalism in these metaphoric and analytical terms as well as those developed earlier in the book. Since the 1990s, it now seems abundantly clear, cities have become more attractive ecological partners than they once were. Whereas cities were once problems, in recent decades they have become solutions—or at least perceived solutions. Where once, as discussed in Chapter 3, cities were the wellsprings of a neo-Malthusian impact crisis, today they are the innovative and creative spaces of a vast reconciliation project, wherein ecology, economy, and equity merge with reenergized democracy to secure and improve future urban worlds—at least in theory. As a consequence, new partnerships both with and between cities all across a shrinking planet have exploded; new arrangements have emerged; new possibilities have appeared.

But what *are* these new partnerships, arrangements, and possibilities? What *are* their recent histories and future geographies? To what extent, moreover, have they been overly romanticized? Alternatively, to what extent (or not) has the global story of cities as new spaces of sustainability now progressed to something more institutionally meaningful and long-lasting? Ultimately, to what extent might this urban romance, *pace* Jane Austin, save our lives?

This chapter engages these questions through select instantiations of green internationalism: (1) the international urbanization of the United Nations, tracing key developments from the Local Agenda 21 (LA21) experience to the recent promulgation of Sustainable Development Goals (SDGs) and the New Urban Agenda; (2) the related but also distinct transnational urbanization of recently constructed, policy-oriented, global city-networks; and finally, (3) the also related but distinct smart city urbanization that has exposed political and social tensions between green corporatism and democratic agency. While these three urbanized forms of green internationalism do not represent the totality of how the geopolitics of global environmental governance are changing, I argue they illustrate the multiple temporalities and shifting spatialities of a post-Westphalian world order struggling to adapt to new challenges.

International Urbanization: The Green United Nations, LA21, Urbanized SDGs

The mid-twentieth century-origins and early institutional, policy, and legal developments of the United Nations (UN) emphasized neither environmental concerns nor cities, much less a global environment infused heavily by the "planetary urbanization" of global economies (Brenner, 2014). Set up in the 1940s as a second and more successful reaction to the repeated horrors of industrialized war, the creation and longevity of the United Nations system is no small achievement, despite a host of contemporary pathologies that suggest "Westphalia is alive, but not well" (Weiss, 2012, p. 19). Environmental concerns were not wholly absent before the 1960s: key UN agencies like the WHO (World Health Organization) and the FAO (Food and Agriculture Organization), for example, had started by then to address global green problems such as fisheries, air pollution, and environmental toxins (Ivanova, 2010). But the environmental turn in the United Nations is most often associated with the early 1970s—and especially with the formal creation and funding of the United Nations Environment Program (UNEP) in 1972.

With the rapid decolonization of Africa and much of Asia still historically fresh, the geopolitics of the UN in the 1970s invariably reflected tensions between the Global North and the Global South as much as the divide forged by the Cold War. Specifically, as Ivanova (2007) suggests, highly developed countries, including the USA and Sweden, worried increasingly about the ecological effects of industrialization (e.g., acid rain) while developing countries were more concerned with the ecological effects of poverty (e.g., sanitation)—a "confrontation" of interests that led to the "compromise" of sustainable development as a normative theory of societal organization defined famously in the mid-1980s by the Brundtland Commission. Ivanova (2007) questions the dominant thesis, propounded by scholars like Speth, Haas, and von Moltke, that the "beginnings of global environmental governance" forged through the difficult creation of UNEP has largely been "a story of purposeful incapacitation" (p. 339). Though historically underfunded, she argues, UNEP has provided crucial leadership in knowledge acquisition and assessment, such as data collection and information exchange; environmental quality management, including goal-setting and crafting agreements; and international supporting actions, focusing on capacity building and technical assistance (p. 346).

Meanwhile, the UN's explicit urban turn—its global policy focus on the management of cities—also came in the 1970s, albeit in the eventual form of UN-Habitat, which was formally established in 1978 in the wake of a major international conference held in Vancouver two years earlier ("Habitat-I"). With origins in the UN's Habitat and Human Settlements Foundation, which was governed briefly under the auspices of UNEP, UN-Habitat also was and remains a Nairobi-based "program" with a comparably small budget rather than an "organization" with the diplomatic clout and legal powers of, say, the World Trade Organization. As Clapp and Dauvergne (2008) observe, there still is no World Environment Organization in this same sense.

Unlike UNEP, UN-Habitat's mandate has been to focus on assisting developing countries with the pragmatic provision of adequate shelter and attendant urban services. Projects support the urban "brown" agenda (Freund, 2001). Running on financial fumes for much of the 1980s, UN-Habitat's fortunes shifted between 1996 and 2002 after the "Habitat II" conference in Istanbul produced the "Habitat Agenda." In 2002, this work expanded as the UN community elevated the role of cities in managing global climate change as well as urban sustainability problems (UN-Habitat, 2012). Finally, "Habitat III," held in Quito in 2016, launched the "New Urban Agenda," a specific response to the UN's overall commitment to Sustainable Development Goals (SGDs), all of which are discussed in more detail below.

Co-located in Nairobi, the originating mandates of UNEP and UN-Habitat—international green and urban brown, respectively—formally coincided in the early 1990s after the Rio Earth Summit around the UN's interagency "Sustainable Cities Program," which has focused since that time on the spaces and governance of the urban environment. For Gebre-Egziabher (2004, p. 65), the Sustainable Cities Program benefitted early on from, on the one hand, "...the extensive operational experience of UN-Habitat in project-level local development and its accumulated expertise in practical urban management," as well as, on the other hand, "...the scientific and technical strengths of UNEP with its data banks, international networks, global accords, and expertise in all aspects of the environment." The Sustainable Cities Program directly supported the post-Istanbul Habitat II Agenda's six-part policy focus on environmental management, social development and poverty alleviation, economic development, shelter, governance, and international cooperation (p. 67).

The operational policy areas of the Sustainable Cities Program complemented the Agenda 21 initiative discussed in the last chapter, by, first, "localizing" (or downscaling) and then, more recently, sharing or "upscaling" program and project experiences around the world. Because Chapter 28 of Agenda 21 emphasized the global role of local governments it is known more popularly as Local Agenda 21 or Local *Action* 21 (LA21). By 2001, ICLEI-Cities for Sustainability counted 6500 LA21 programs in 161 countries, even as a uniform and accepted definition of LA21 remains elusive (see: https://www.gdrc.org/sustdev/concepts/18-la21.html). While 6500 sounds impressive in isolation, there are more than 19,000 general-purpose municipalities in the USA alone. The *uptake geographies* of LA21, then, are radically incomplete—even as discussion of LA21 remains remarkably robust a generation later.

Intellectual geographies of work on LA21, for example, serve as a useful (if imperfect) proxy for wider patterns (Fig. 4.1). Google Scholar

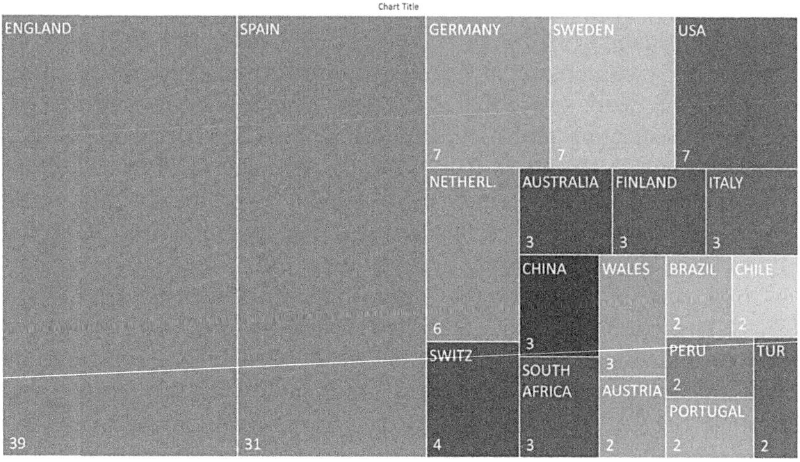

Fig. 4.1 Intellectual map of LA21 research (*Chart based on data derived from* Web of Science Core Collection, Topic = "Local Agenda 21," 1995–2017, sorted by geography/region (3/1/2018))

generated 16,200 sources in February 2018 for "Local Agenda 21." The Web of Science cited 132 scholarly publications between 1997 and 2018. Half this work is either in/on the UK or Spain. Western European countries like Germany, Sweden, Finland, the Netherlands, and Italy making up another 25% of papers. Although the American and Chinese economies together constitute 42% of global GDP, LA21 has figured less in either country's research agenda, suggesting other "frames" for the localization of global sustainability, such as smart growth in the USA (Krueger, 2010).

The spatialities of LA21 are theoretical and empirical. They relate to the ways in which local institutions effectively restructure both procedural and substantive concerns around the shared sustainability discourses of the wider green international project. Barrutia, Echebarria, Paredes, Hartmann, and Apaolaza (2015) theorize LA21 as a "participatory reform" to older and more familiar, strategic-rational planning models of governance, which always emphasize long-term thinking and interrelated comprehensiveness. Ideally, they argue, interscalar political commitments to LA21 programs—which involve diplomatically engaged states, empowered regions, and/or willing and well-capacitated localities—enhance policy legitimacy; reduce conflicts; imaginatively fit contingencies to global problems; empower (now) more and better educated citizens; and, not least, institutionalize the bureaucratic use of performance indicators. However, actually existing experiences with LA21—material realities across communities—are still poorly understood. We know for certain that LA21 has flourished across Western Europe, in general, and countries like Britain, Spain, and Sweden, in particular. But what does such flourishing really mean?

Theoretically informed case study work on the British, Spanish, and German experiences provides initial insights into this question, without necessarily allowing anyone to generalize these experiences and insights to other world regions and urban areas absent considerable care and appropriate caution (Barrutia & Echebarria, 2013; Garcia-Sanchez & Prado-Lorenzo, 2008; Jonas, While, & Gibbs, 2004; Llamas-Sanchez, Munoz-Fernandez, & Maraver-Tarifa, 2011). Western European experiences can be usefully contrasted, however, with different analyses of how the relatively recent policy mergers of UNEP and post-Istanbul UN-Habitat (III), and specifically LA21 and the Sustainable Cities Program, have influenced and shaped, for instance, sustainable urban development experiences in Nigeria, South Africa, and other African cases (Ogbazi, 2013; Roberts & Diederichs, 2002).

Although the Brexit vote has changed governance dynamics within the UK in ways we are only now just appreciating, the rich literature on the spatialities of the LA21 experience in the late 1990s and early twenty-first century highlights, in my view, the theoretical importance of "the sliding scale of spaces" addressed in the first three chapters. The popularity of LA21 in the UK—its rhetorical uptake, institutional rollout, and territorial effects—foregrounds more than just a fragmented and uncoordinated empirics of "radical local environmentalism that exists independently and outside other pressures" (Jonas et al., 2004, p. 153). It highlights how different scales of government and governance both within and beyond the UK adjusted to shifts and pressures associated with the contradictions and concerns attending both neoliberal capitalism and global unsustainability. Deploying neo-Gramscian state theory,[1]

[1]An extended treatment of this theoretical approach, especially as influenced by Bob Jessop, lies beyond the present book. Suffice to note only that neo-Gramscian state theory, as MacLeod and Goodwin (1999, p. 515) put it, directly citing Bob Jessop's work, explores "how political, intellectual and moral leadership is 'mediated through a complex

Jonas et al. contend that LA21 in the UK reflected how the British state, particularly during the New Labour era, strategically and selectively operationalized "ecological modernization." That said, like any state, the British state "...is not a 'spaceless' and 'place-indifferent' entity," so it also cannot "organise from above" without detailed and ongoing policy and regulatory attention to local variabilities in power, interests, values, problems, and capacities (p. 155).

Jonas et al. (2004) offer a range of relevant insights into early cases of LA21 implementation: Manchester, Leeds, and Lancashire County. Despite their close proximity, each case represents what they call a different "trajectory" in how LA21 processes shaped *and were shaped by* wider pressures and transformations occurring at different scales. Lancashire County, for example, was one of only fourteen local authorities in the world that participated in the "Local Agenda 21 Model Communities Programme," a post-Rio initiative supported by the International Development Research Centre (IDRC), the United Nations Development Programme (UNDP), and the Netherlands Ministry of Foreign Affairs. The purpose of this initiative was "to aid local governments in implementing Chapter 28 of Agenda 21, the global action plan for sustainable development" (see: http://archive.iclei.org/index.php?id=1202).

In a self-assessment of their own work, the Council emphasized the still voluntary (but politically advisable) incorporation of broader LA21 goals and ethics into its statutory requirements to prepare a "Community Strategy"—touting progress in "projects" such as promulgating an environmental purchasing policy and improving access to the countryside by "public transport, walking, cycling and horse riding" (Lancashire County Council, 2003a, p. 4). A separate end of term report similarly concluded that, "Overall good progress has been made in implementing performance measures in Year 2. [:...] 41% of performance measures have been achieved, 48% are in progress and only 3% have made no significant progress" (Lancashire County Council, 2003b,

ensemble of institutions, organizations and forces operating within, orientated toward or located at a distance from the juridico-political state apparatus.'" Their work is important here because it accounts for major changes in the territorialities and relationalities of the state without accepting *tout court* the "denationalization of statehood or the hollowing-out of the state" (ibid.).

p. 4). Jonas et al. (2004) nonetheless suggest—less sanguinely—that Lancashire's earlier leadership in environmental policy was "compromised by state modernization" (p. 164). The spatialities of LA21 in Britain, the uneven urbanization(s) of an internationally negotiated kind of green internationalism, necessarily reflect a complex politics of scale—evolving external (diplomatic) and internal (territorial) statecraft as well as localized policy agency and networking capacities (MacLeod & Goodwin, 1999). Ironically, the widespread uptake of LA21 in Britain may have had more to do with the enervation of local democracy under Thatcher as with the presence of institutions looking to implement a new phase in international environmentalism.

Localized policy agency and networking capacities also constitute a major theme in work on the LA21 experience in Spain, especially with respect to the Barcelona area and the Basque County/Autonomous Region—both of which have long generated profound challenges to the precarious territorialities of the Spanish state. Barrutia and Echebarria (2013) suggest, for instance, that "successful" LA21 experiences should not really be considered the sole responsibilities of municipalities but instead should be examined as transcendent networks, particularly as regional complexes between interacting municipaliites. In their judgment,

> diverse empirical evidence regarding LA21 seems to indicate that LA21 dissemination tends to be higher in territories where networks have emerged.... Municipal governments participating in neworks appear to obtain important benefits from the transmission of experiences and inter-municipal collaboration. (p. 43)

Collaboration is the dream condition of governance scholars and practitioners, the elusive Holy Grail that believers (and even a few skeptics) search for in the romantic hope of territorial salvation (Thibert, 2015). Close substitutes or key corollaries often include participation, involvement, empowerment, synergies, transparency, "joined up" thinking, etc. In planning theory, the successful *routinization* of collaborative cultures and discursive habits generates what Patsty Healey (2006) has called "relational resources." In geography and economic governance, collaboration potentially begets improved institutional thickness and the possibility of building new kinds of city-regions that actually learn how to innovate (OECD, 2002).

Alternatively, city-regional collaboration is *itself* a form of innovation as self-organizing, market-driven interactions per se are otherwise hindered by inadequate institutional and political conditions. Albeit crucial to "region-building" in a global economy, private interactions seldom generate sustainable and/or politically acceptable development trajectories *in and of themselves* (Meadowcroft, 2011). Market-based interactions are, as Charles Lindblom (1977) memorably phrased it, all fingers but no thumbs. Competing political claims for scarce resources place constant pressures on public authorities at various scales to "use their thumbs." Public authorities in nominally democratic societies that need legitimacy must address a host of extra-economic problems: viz. uneven spatial development, cultural and demographic change, concerns with accountability and fairness, ethnic and racial tensions, market failures such as negative externalities and the undersupply of public goods—and finally, unsustainability broadly understood in its economic, social, and ecological dimensions. More generally, collaboration requires institutions that are able to absorb and/or mitigate the transaction costs of territorial management. Ronald Coase has aptly summarized this last challenge:

> The economic system is extremely complex. You have ... differentiated firms and narrowly specialized firms, vertically integrated firms and single-stage firms; you have in addition non-profit organisations and government entities – and all bound together, all operating to form the total system. [The key therefore is] the comparison between the additional production resulting from the rearrangement of activities and the cost of the transactions needed to bring the rearrangement about. However, the transaction costs depend on the working legal system. They also depend on the political system, they depend on the educational system, and they are interrelated with the other social systems. (cited in Bertolini & Giovannetti, 2006)

Barrutia, Aguado, and Echebarria (2007) discuss the local implementation of Agenda 21 in the Basque Country/Autonomous Region, situating these experiences in the context of broader implementation problems, such as the tendency to see sustainability as an environmental rather than also social and economic concept as well as the tendency to make plans rather than deliver programs and projects. They highlight common LA21 problems documented elsewhere, like weak stakeholder participation and insufficiently developed links with local problems. Within the global A21 framework, they explore how local authorities there have emerged to set up actions reflecting a scaled-down proximity to people. At the same time, given significant resource constraints within most local municipalities, they also identify and theorize the (scaled-up) institutionalization of a "successful policy network," describing it as,

> a process of public-private collaboration within which the regional government provides leadership, motivation, knowledge resources and an ability to pull all the stakeholders together, and in which councils, led by their Mayors, work as a network, saving resources and sharing knowledge and best practices. (p. 34)

In their later work on the subject, Barrutia and Echebarria (2011) explore whether the embrace of LA21 in this region represents a broad but weak diffusion of innovation, arguing that actors *who work together* "cocreate" capacities that matter for local sustainability policies.

As international green discourses like LA21 circulated and gained adherents, the British and Spanish cases illustrate that Europe is both a continent of relatively strong nation-states but also of distinctive cultural regions. LA21's generically popular but uneven implementation within Europe illustrates how discourses originating in the UN system cascaded down and diffused through the intertwined scales and variegated spaces of diplomatic, national, and territorial policy-making. Across Germany, for example, LA21 programs started a bit later than in the UK and Sweden, largely peaked between 1997 and 2002, and seem to have left behind a specific geographical imprint reflective of the federal political system as well as local socioeconomic and institutional conditions, including motivations and capacities to collaborate. In their important study, Kern, Koll, and Schophaus (2007) noted that by 2006 nearly two-thirds of local authorities had adopted LA21 programs in Hesse, situated in the central-western Germany, but only 1.8% in Saxony-Anhalt, located in former East Germany, had done so. Still more interesting for present themes, they noted that all three of the county's "city-states"—Berlin, Hamburg, and Bremen—were early policy pioneers and that, more generally, "...99% of all cities with more than 100,000 inhabitants and 94% of all cities with more than 50,000 inhabitants have initiated a LA21 process" (p. 610). The green urbanization of the LA21 initiatives across Germany, in other words, means more than just a localization of UN agreements; it literally highlights the elevated role of larger cities and urban-regions in global environmental politics in a specific state structural context.

Over the course of the 2000s and early 2010s, LA21's initial localization and urbanization of the UN's overall approach to a new green internationalism steadily merged with other initiatives, including annual meetings of the Conference of Parties (CoP) around global climate action, and especially in the wake of the Habitat-III conference in 2016 the promulgation of Sustainable Development Goals (SDGs). The relationship between global climate action and cities is discussed in the next section. The promulgation of SDGs, though, warrants a bit more treatment here, especially as these goals relate to the so-called "New Urban Agenda," which probably represents the most significant evidence yet that cities have arrived in green geopolitics.

The New Urban Agenda (NUA), adopted by the UN General Assembly in December 2016, reflects how the international community increasingly perceives urban systems and the physical forms of urban space as "powerful tool[s] for sustainable development for both developing and developed countries" (United Nations, 2017). Both systems and spaces, the NUA holds, should reflect "the science of cities," serving "standards and principles" for planning, construction, development, and management. Such standards and principles aim to link the NUA with the 2030 Agenda for Sustainable Development, notably Goal 11 to make cities and human settlements inclusive, safe, resilient and sustainable, along with the Addis Ababa Action Agenda of the Third International Conference on Financing for Development and, once again, the Paris Climate Change Agreement. The New Urban Agenda perfectly represents the international urbanization not only of national, subnational, and local governments, but also of the World Assembly of Local and Regional Governments, which is focused on decentralization, subsidiarity, and integrated territorial governance.[2] Notably, the agenda reflects empirical efforts by various institutions to incorporate pragmatically Henri Lefebvre's theoretical concept of the "right to the city" in legislation, political declarations, and charters (United Nations, 2017).

[2]The World Assembly of Local and Regional Governments, founded in 2013, is a global action forum that "coordinates major international networks of local governments to undertake joint advocacy work relating to global policy processes." Its members include many of the most influential and powerful transnational municipal networks formed to address global problems in recent years, including UCLG, ICLEI, C40, and Metropolis, among many others. See: https://www.global-taskforce.org/about-us.

Taken at face value, the UN's NUA embodies a syncretic admixture of urban ideas and normative claims, from the rational-technical language of scientific systems to the radical-normative activism of Lefevrean social theory. The system-*cum*-spatialities preferred include, for instance, a commitment to "a smart-city approach that makes use of opportunities from digitalization" (p. 19)—dealt with later in this chapter—no less than "gender equality and the empowerment of all women and girls" (p. 3). Strong and repeated emphasis is placed on "inclusionary" and "participatory" processes, but also "standards and principles" of spatial place-making that arguably embody extra-local norms of green city building which now circulate through global space and pre-shape how local problems are framed and discussed (e.g., compactness and density, polycentrism, mixed uses, infill, planned urban extension strategies, etc.)—a dynamic Lefebvre (1991) elsewhere theorized as the "spaces of representation" central to the planned production of abstract space-economies in late capitalism.

More generally, however, it is unclear how the NUA and Agenda 2030 improve upon previous efforts by UNEP and UN-Habitat, including LA21 and the Sustainable Cities Program, particularly in societies with locally weak or at least constrained institutional capacities in spatial planning and policy development (Ogbazi, 2013). As might be expected, urban scholars of various kinds are now starting to push back against the synoptic claim, emerging forcefully over the past ten years or so, that cities or even collaborative urban regions can really play a truly decisive role in world politics and the pragmatic management of global problems absent more comprehensive changes in the territorial structures of politics and policy-making (Toly, 2017).

The legal, constitutional, diplomatic, and financial constraints placed upon the actually existing power of cities—and the variety of economic and cultural forces that work against city-regional creativity—are already well known (Frug, 2008). At a minimum, cities around the world are still embedded within "nested hierarchies" (Hill, 2004). As cities network relationally with one another across these "nests," both the potentials and pathologies of multi-level territorialities modify global actions. "In the absence of a tectonic shift in individual identity, collective security and political sovereignty from nation-states to a world government," Hill writes, "globalisation will continue to unfold by and through nation-states" (p. 382). Following insights originally suggested by Jane Jacobs, a host of extraordinary cities—Catal Huyuk, Ur, Babylon,

Athens, Rome, Constantinople, Xian, Tenochtitlan, Florence, New York, etc.—may have long constituted the well-springs of wider patterns in world development, as Peter Taylor (2012) and others show (Soja, 2000); but the more recent "celebration" of the city, Robert Beauregard (2018) counter-argues, is now "hyperbolic." Cities are endlessly creative places, without question, but also "cauldrons." They are spaces of contradiction, not just spaces of problem-solving.

Case Study: The creativity and cauldron of Cape Town. The creativity, contradiction, and cauldron constituting the cities that Beauregard highlights in his US-based treatment are nonetheless nowhere more on display than in metropolitan South Africa. In the post-apartheid era, South African urbanism is especially instructive to our themes here because major (South) African cities like Johannesburg, Cape Town, and Durban *have* been institutionally empowered through significant reforms in at least two major ways. First, South Africa's Constitution recognizes local governments as distinctive spheres of government, rather than, for example, legal creatures of the state (as in the USA); second, in three major stages over the course of the 1990s, larger globalizing city-regions like Johannesburg, Cape Town and Durban also underwent a comparatively remarkable process of formal political amalgamation, becoming self-styled "unicities" at the turn of new millennium (Jenkins & Wilkinson, 2002; OECD, 2008; Watson, 2002). Put another way, their "nests" *were* padded by a reformed state apparatus.

While Johannesburg has focused on consolidating its already strong economic and financial position as the African continent's most important city, particularly in regard to facilitating foreign and especially Chinese investment (Rogerson, 2004; Rogerson & Rogerson, 2015), the creativity, contradiction, and cauldron in how green internationalism is urbanized within South Africa is arguably more pronounced in Cape Town.

Profoundly beautiful and endlessly alluring, it is rather easy to fall in love with Cape Town, but perhaps just as easy to romanticize its geo-histories and socio-spatial possibilities (Fig. 4.2). Cape Town, as almost everyone who cares about cities and global environmental politics knows, has been experiencing the reality of global climate change *now*, not as a distant possibility for future generations mentioned abstractly by the Brundtland Commission in the 1980s. Located in a mild Mediterranean-like climate zone, drought-stricken Cape Town is today (c. 2018) essentially the first major international city to face the absence of a reliable

Fig. 4.2 Cape Town from Table Mountain in 2014 (*credit* Yonn Dierwechter)

water supply. Cape Town has given the world a new term: "Day Zero"—the day water runs out (Van Loon, 2018).

Cape Town, like everywhere, is manifestly unsustainable. Its liquid ecologies cannot reproduce its long-racialized (apartheid-induced) sociologies, uneven (human) geographies, and dual ((in)formalized) economies—much less help to transform justly these inherited sociologies, geographies, and economies. The social contradictions that stoke this city's cauldron won't be cooled in the least by the absence of water. Yet as this city searches for long-term solutions, the world will watch closely. Urban water crises will grow both more common and more geopolitical.

Invariably, possible de-salinization processes contemplated to get by this time round, or during the inevitable *next* drought, could potentially create new forms of "planetary urbanization" (Brenner & Schmid, 2011): the Cape of Good Hope's immediate coastlines could be pulled into the city-making process in novel ways. The ocean is not a city; but the ocean is and could be urbanized, harvested, and reconstructed as a socio-technical extension—a neighborhood—in a new world of unsustainable metabolisms. Surrounded by water, a waterless Cape Town would at minimum invent new forms of sustainability policies, new experiments in political management, new challenges for global learning, and new geographies of the urban condition.

So Cape Town must also be—as it often has been in the past—manifestly creative. While theories of urban creativity suggest the importance of *endogenous* innovations (Törnqvist, 2004), including, as discussed earlier, the political capacity to coordinate diverse actors across city-regional policy complexes (Perlman & Jimenez, 2010), actively learning and appropriately borrowing from others is also important. Cape Town, for example, might learn more from Melbourne, whose own experiences with severe drought between 2001 and 2010 led to important water conservation reforms (other than engineering fixes) that have lasted beyond the drought; it might also continue to learn from similarly challenged cities in Brazil, a country with a high Gini coefficient long compared analytically with South Africa (Watson, 1991).

In the 1990s, Cape Town explored Curitiba's well-known experiences with bus rapid transport (BRT), eventually launching its own BRT system, now known as MyCiti. These transit shifts have been part on an overall development framework that sought—and still seeks—to emphasize engaged, empowering, participatory planning, and socio-spatial cohesion through what the South Africans have long called Integrated Development Planning for Developmental Local Governance (Pycroft, 1998; Sowman & Brown, 2006). Both these concepts—Integrated Development Planning and Developmental Local Governance—also reflect early responses to LA21 and global Sustainable Cities discourses, in Cape Town as much as anywhere. Venter (2007, p. 93) noted a decade ago that the philosophies, principles, and process of Integrated Development Planning "are consistent with Local Agenda 21." They did not emerge *because* of LA21, but instead intermingled with international diplomatic frameworks that increasingly defined urban policy-making after the end of the Cold War.

In addition, Cape Town's socio-spatial vision for sustainable urban development, like the New Urban Agenda and Sustainable Development Goals, still emphasizes "inclusionary" and "participatory" processes but also "standards and principles" for spatial place-making (City of Cape Town, 2012). South Africa was instrumental in elevating the international importance of the informal city in Habitat III's New Urban Agenda. The premier of Gauteng, in particular, argued publically that informal settlements, where an estimated one in seven people on the planet now lives, should be a central to a "new urban vision in line with the United Nations' Sustainable Development Goals adopted in 2015" (Bosworth, 2016).

Discussion of the preferred form and function of the post-apartheid city-region has long championed a strong and well-developed critique of the apartheid city as well as concerns with unsustainable human settlements associated with modernist-era urbanization and, as discussed earlier, a global policy convergence around new green growth management theories (Dewar, 1979, 1998; Dewar & Watson, 1990). During the 1990s and beyond, local planning theory and to some extent metropolitan spatial policy emphasized the urgent need to restructure the region functionally over time according to a set of relatively simple but powerful spatial development principles that were concomitantly anti-apartheid and anti-modernist—if physicalist:

- "Compaction as opposed to sprawl;
- Integration as opposed to fragmentation;
- Greater structural complexity;
- Mix, not separation;
- A fixed, permanent urban edge; and;
- Place-making and resource creation" (Dewar & Todeschini, 1998, p. 9).

The early impact of these general principles was expressed in municipal and regional spatial planning policy and discourses for several years after the formal end of apartheid. But pilot projects that actually attempted to localize some of these general principles of town-building—such as the Wetton-Landsdowne Corridor Initiative in the impoverished Southeast of the city, itself also directly inspired by Curitiba (Cape Town City Council, 1998)—quickly exposed how the local geopolitical economy would consistently enervate rather than edify this remarkably progressive, African-based, planning-shaped, model of sustainable urban development (Watson, 2002).

Rather than the development patterns broadly consistent with LA21 values, Sustainable Cities, and the New Urban Agenda, the continuation of powerful centrifugal forces and especially the launch of new satellite proposals—even whole new towns—have threatened Cape Town's political search for a more sustainable city-regionalism integrated justly with the global economy. The post-apartheid projects of social integration and economic justice have struggled to surface from the heavyweights of both extant regulatory inertia and path-dependent investment interests. Economic regeneration is most needed in the impoverished and fast-growing "southeast" quadrant of the city-region, near African Khayelitsha and Cape Colored Mitchell's Plain. No one disputes that. Yet the Minister of Local Government, Environmental Affairs and Development Planning nonetheless approved in 2014 a provisional scheme to develop "WesCape," located some 20 kilometers North-by-Northwest of the Cape Town central business district.

At hypothetical "build-out" in the coming decades, WesCape would include 800,000 people, 200,000 homes, 400 schools, 90 community facilities—even a university. This is rather like placing the entire city of San Francisco *outside* of Cape Town rather than within it. For its advocates, such developments deliver global sustainability through private-sector led self-containment, particularly when linked up with "smart" technologies. Vanessa Watson, Nancy Odendaal, and other local experts, however, believe these colossal, mega-project approaches to development badly abrogate key principles of the Cape Town Spatial Development Framework, adopted officially after extensive consultation in 2012 as part of the city's overall Integrated Development Plan (City of Cape Town, 2012). For these observers especially, Cape Town is thus engaging in "a complete planning disaster" that betrays democratically and organically developed visions of the just and sustainable post-apartheid city (Watson, quoted in Davis, 2013).

While critiques of Cape Town's post-apartheid development emphasize the unjust unsustainability of low-density, disconnected, segregated, oil-dependent landscapes, local discourses, mirroring the New Urban Agenda's concerns with smart cities, increasingly counter-suggest more broadly that "data is the new oil" (City of Cape Town, 2016, p. 12). Here Cape Town is now on an exciting if difficult "digital journey," guided somewhat ironically by the Constitution and the same (abrogated) Integrated Development Plan discussed earlier, to build "digital institutions...driven by data and evidence, focusing relentlessly on [its] customers," according to the Executive Mayor (ibid.). Smartness, so conceptualized, is a multi-dimensional effort to move largely corporate sector innovations in digitization, illustrated best by Amazon, Netflix, Uber, and Airbnb, to public sector values and modes of "business" (p. 9).

Interestingly, as Herrschel and Dierwechter (2018) have shown, the origins of Cape Town's digital journey largely coincide with the national creation of formal Unicity authorities to redress the pernicious legacies of apartheid planning and racialized administration, merging pressing questions of city-regionalism with parallel concerns around smart governance, smart economies, smart infrastructure, smart citizens, and so on. Steady "digitization" has meant, however, that smart governance, for example, is not simply about maximizing the number of permit, license, and service applications issued *electronically*, but also about urban-political efforts "to lobby for changes to national legislation and regulation which stifle local innovation" (p. 14).

This last point reminds us that the urbanization of green internationalism is difficult to explain much less advance without parallel attention paid to the shifting territorialities and potential competencies and policy responsibilities of the formal nation-state and interstate apparatus, however legitimately reviled or seemingly broken.

Ideally, the ***kind of state*** that cities engaged in global environmental questions now most need to help transition toward a more robust sustainability simply does not (yet?) exist in contemporary South Africa—nor most anywhere else either. Following the end of the "post-apartheid" era, marked for convenience here by the death of Nelson Mandela in 2013, the South African state rapidly regressed and decayed under the outwardly corrupt presidency of Jacob Zuma (Johnson, 2015). The emergence of a "developmental state" focused on neo-Keynesian redistribution was curtailed by 1996 when the African National Congress formally abandoned its initial Reconstruction and Development Programme (Ansari, 2017). But the neo-patrimonial, rent-seeking state under Zuma went far beyond otherwise now familiar concerns with an early "neoliberal" turn. Prominent scholars working on the State Capacity Research Project (2017) have argued that South Africa under Zuma actually experienced a "silent coup."

The international urbanism discussed here, the "greened up" urbanization of an inter-state system whose ideas, values, and programs still flow daily through UN agencies and other Westphalian bodies, suffers from such dysfunctional dynamics. For this reason, leading figures like Ben Barber (2013) understandably sought to change the conversation entirely, emphasizing how cities in new international forms of (green) agency could change how policy action might and increasingly do relate on the ground. His provocative call for a "global parliament of mayors" at least offers a different diplomatic imaginary—albeit one that romanticizes the city and too quickly cuts it off from other scales of (em)power(ment) and "puzzling," to use Hugh Heclo's terminology. As we have just seen in the case of how the UN interacts increasingly with bodies like the World Assembly of Local and Regional Governments, a new "transnational urbanism" has indeed emerged. The next section discusses this as a related, but also distinctive, manifestation of the recent urbanization of green internationalism.

Transnational Urbanization: The Explosion of Green Policy City-Networks

The international perspective just highlighted reflects how the urbanization of various states has forged new kinds of formal diplomatic agency between countries in the interstate system as these are reflected in recent UN policy developments like the New Urban Agenda. Critics of that system consider it outmoded—essentially a lumbering form of top-down governance predicted on a Westphalian order that has been crumbling for decades. New forms of transnational urbanism, in contrast, supposedly emphasize the possibilities of bottom-up transformation (Bond, 2010). The most prominent of these new forms of agency are policy-oriented city-networks focused on global-scale problems like climate change and adaptation, such as Climate Alliance and the Covenant of Mayors. In turn, many of these still relatively new city-networks, especially the high-profile C40 club, typically transgress national borders and state-based systems and are animated politically by strong and/or charismatic mayors who have breached the borders of national political cultures (e.g., figures like Ken Livingston, Michael Bloomberg, Anne Hidalgo).

Michael Peter Smith (2001) first developed the concept of transnational urbanism with immediate respect to the variegated localizations of globalization. More recent work has reflected the explosion of green policy city-networks over the past two decades or so (Castán Broto & Bulkeley, 2013; Keiner & Kim, 2007; Kern & Bulkeley, 2009). Various authors have offered related though distinctive conceptual framings and spatial imaginaries that similarly seek to characterize an emerging, arguably "post-Westphalian" geographic space, such as Liam Magee's (2016) textured metaphor of a new world order shaped by "interwoven cities"—developed for this book series—as well as Henner Busch's (Busch, 2015, 2018) recent arguments around the emergence of "entangled cities" focused on global climate action within and beyond Germany. For Busch (2015, p. 3), entangled cities are "institutionalised spaces *where local governments from different countries* come together as equitable partners in an exchange on climate change related issues" (p. 3, emphasis added).

Claims of "interweaving" and "entanglement" suggest, in my view, a level of functional but not legal or political integration that goes increasingly beyond the otherwise voluntary and non-hierarchical basis for the formation and development of transnational spaces of green internationalism. The impacts of membership are still insufficiently understood (though see, e.g., Krause, 2012). Yet recent work does suggest that transnational municipal networks—Busch's "entangled cities"—demonsratably influence policy adoption in a range of key arenas, including energy, finance, outdoor lighting, transport procurement, land use regulations, and waste management (Rashidi & Patt, 2018).

Not all city-policy networks focused on planning for climate change are *transnational* municipal networks. Primarily *national* or regional networks, such as the Mayors Climate Protection Agreement within the USA, provide one such example (Dierwechter, 2010; Dierwechter & Wessells, 2013). Nor are all green city-policy networks *transcontinental*, integrating cities in common projects across world regions. Key TMNs in Europe include the Alliance of the Alps, formed in 1997. This group focuses on smaller communities that shape the Alps eco-region. The Alliance of the Alps is interesting here because it was forged in response to the 1991 Alpine Convention, "an international treaty between the Alpine Countries (Austria, France, Germany, Italy, Liechtenstein, Monaco, Slovenia and Switzerland) as well as the European Union, for the sustainable development and protection of the Alps" (see: http://www.alpconv.org/en/convention/history/default.html). Its original construction and maintenance are multi-scalar, drawing on the diplomacy and active agency of member states and supra-national institutions as well as local municipalities and various civic partners. Del Biaggio (2011) notes that "the influence of the national and supranational levels can be seen, for example, in the organisational structures of the networks and in the composition of the steering committees." In point of fact, the "multi-scalarity" or "multi-level" nature of otherwise "urban" climate action is now a relatively well-developed theme in the wider literature (Bulkeley, 2005; Kern & Bulkeley, 2009).

Multi-level influence does not necessarily mean efficacy. City-networks of relatively small communities who rely on state support are different from transnational municipal networks built around and to a large degree by major political and economic players like London, New York, and Barcelona. The novel green internationalism recently forged by much-discussed networks like C40 Cities, headquartered in New York, and, to some extent, United Cities and Local Governments (UCLG), headquartered in Barcelona, also suggests geographical unevenness and hierarchies of power both *within* and *across* an otherwise vibrant ecosystem of urban experimentation. Put another way, networks of global policy action—albeit new forms of "clubbing together"—may be "overriding if not effacing an earlier core-periphery system of territorialized states" and their international organizations like UNEP and UN-Habitat (Scott, 2001); as such they nonetheless reflect their own distinctive cartographies of economic, social, and political development and thus their own power-geometries of urban political and international environmental affairs.

Such power-geometries, which Doreen Massey (1993) first discussed as the time–space compressions of globalization, demonstrate how forms of horizontal networks are still shaped exogenously by various kinds of traditional structures associated with states. Setzer (2015) shows, for example, that the diplomatic activities of both local and regional governments are constrained by international and domestic legal frameworks of all kinds. In his 2018 state budget, as one recent example of this argument, the Democratic governor of New York, a powerful national figure since at least the gubernatorial administration of Franklin Delano Roosevelt, not only thrust a "defiant chin" upwards to President Trump, a Republican; he also asserted a "heavy hand" downwards, to the mayor of New York City, a fellow Democrat. The governor's budget provided for more state-level oversight of New York City's public school system; more state involvement in the city's public housing authority; and a more aggressive use of the state's legal right to develop Mid-Town

Manhattan around Penn Station. Schools, housing, and real estate development are the bread-and-butter of US "urban" politics. One New York City Assemblyman, another nonplussed Democratic, hyperbolically but not without cause called the governor's moves the death of home rule (Wang & McKinley, 2018). Relatedly, it is hard to see how mayors can—and increasingly do—rule the world if they obviously can't even rule their own city in the most basic of policy arenas. Territorialized power-geometries still matter. Multi-scalar state systems still matter.

On the one hand, scholars increasingly recognize that transnational networks of subnational actors foster intercity and international policy learning and change and, at least in some policy arenas in some cities some of the time, actually help local governments to adopt and pragmatically carry out important activities like greenhouse gas reductions (Lee, 2013; Lee & van de Meene, 2012). That is an important insight. It helps to sustain the imagined possibilities of a budding and quite real romancing of cities. On the other hand, new green urban networks may well replicate (and possibly, if not always, *reinforce*) societal path-dependences associated with extant power formations (Acuto & Rayner, 2016), perhaps even including the core/periphery/semi-periphery world system theorized originally by leading figures like Immanuel Wallerstein (1979) in the 1960s and seventies.

If this once-influential approach in development studies and international political economy now seems antiquated by the empirical transformations wrought by neoliberal globalization and post-modern attacks on grand metanarratives, it is sobering to consider, with Van Hamme and Pion (2012), the perhaps inconvenient thesis that "core countries are still characterized by the intensity of their reciprocal relations while peripheral countries have few economic relations with a limited number of core countries [demonstrating] that [the] core–periphery division of the world has not lost its pertinence" (p. 65). In simpler language, France is more important to Cameroon than Cameroon is to France (p. 70). Global networks have not "flattened" the world but instead have created "new mountains" (Rodríguez-Pose & Crescenzi, 2008). Van Hamme and Pion relatedly critique network analyses of interfirm relationships that abstract nodes and links from what they call the "old structures" associated with territorialized forms of uneven accumulation and geoeconomic power. Methodologically, they use "*k*-core" graphing techniques generated from data on foreign direct investments and trade flows to identify those "regions" of network flows and interrelationships characterized by "pockets of densely connected actors" (p. 69).[3]

Interestingly, the transposition of *k*-core analyses from interfirm and trade linkages to the policy spaces of TMN's who are now collectively focused on global climate action reveals similar patterns of "densely connected actors." Using a sample of thirteen TMNs and 7677 cities and regions as "nodes," Bansard, Pattberg, and Widerberg (2017) show that only a relatively small group of 288 cities and regions are actually part of *more than one network*; in addition, an even smaller subset of 45 nodes connects together all thirteen TMNs they study (e.g., Climate Alliance of European Cities, Covenant of Mayors, C40 cities, EnergyCities, World Mayors Council on Climate Change, etc.). Offering their own geographical vocabulary, they call these highly connected nodes "super-connectors" and argue, in particular, that superconductors are mostly European cities, such as Barcelona, as well as North American states like California. (Analogously, Barcelona is more important to other networked cities than these cities are to Barcelona.)

Noah Toly (2008) is correct to ask if cities can ultimately help to redefine the rules of global environmental politics by developing and taking a stand on inefficient norms associated with a climate-stable future. He and others are correct to explore just how such networks might bypass nation states, helping to apply normative pressures from multiple levels and directions. But as Bansard et al. (2017, p. 230) also observe, the "miracle of civic glocality" envisioned by leading champions of urban sovereignty like Ben Barber (2017) is still profoundly *territorialized*, embedded deeply inside the geo-histories of a world space-economy freighted with competitive interests no less than cooperative values.

Despite these realities, participation in TMNs, which is basically voluntary and always provisional, continues to grow and expand. Why, exactly? What does it mean?

For Lee and van de Meene (2012), networks, however unevenly constituted over world space, are nonetheless also characterized by "teaching" and "learning." It is relatively easy to appreciate why the promises of "learning" might entice cities to join specific transnational networks,

[3]According to Dorogovtsev, Goltsev, and Mendes (2006) *k*-core techniques extract and index interconnected parts of complex networks—communities, cliques, cores, etc.—in order to locate relations between substructures, which in turn illuminates the topologies of real-world networks.

even as resource limitations may also place major obstacles on how many networks they might join and how active and frequent they might participate. While "learning" might not require network membership, but instead ad hoc, bilateral exchanges between cities (like Curitiba and Cape Town discussed earlier), the professional routinization of network membership appears to explain at least some of their initial development and popularity. But what about "teaching"? Assuming "superconductors" are also "teachers," why engage so heavily in the genesis and maintenance of TMNs? What motivates "teachers" to imagine, built, recruit, join, and steadily enrich these new kinds of policy space, especially if the material benefits are (at best) still largely theoretical and normative rather than decisively proven?

Tensions and frustrations with higher-scale authorities, especially national or federal institutions, provide a possible answer to this key question. Existential fears about the costs and consequences of rising sea levels or, as in Cape Town, waterless populations, provide a second possibility. The political culture and long-standing ideological heritage of self-consciously "progressive" cities provide still another perspective on both motivations and expectations. In the literature on the USA, Zahran, Brody, Vedlitz, Grover, and Miller (2008) argue that "civic capacity" better explains why, for instance, American cities tend to join networks like ICLIE's Cities for Climate Protection Program than perception of "risk" or actually-existing "stresses." But why mobilize that capacity in the first place? In one analysis of the metropolitan geographies of US climate action, Dierwechter (2010) focuses on the importance of *local economic structure*, marshaling empirical support in the US context for While, Jonas, and Gibbs's (2004) theoretical concept of a "sustainability fix," applied originally in the British context. For Dierwechter, *central* cities (like Seattle, Miami, Atlanta, or Portland) and to a lesser extent edge cities that are also rich in jobs constitute notable exceptions to an otherwise vast sea of suburban inaction, a theme also observed and further developed by other US-based researchers (Osofsky, 2005, 2015). As Dierwechter puts it,

> [Certain] central cities [in the USA] ... have taken on climate change, but vast swaths of metropolitan suburbia continue to reproduce a political geography of local free-riding. While any number of reasons might explain this hypothesized 'local divide', the discussion in this paper has focused ultimately on a modified form of [the] concept of a 'sustainability fix'.

> This concept draws attention to the basis of ongoing wealth creation in any given community. Though originally applied to British cities, the concept of a 'fix' appears to have theoretical merit in the metropolitan American context. (p. 79)

For Dierwechter (2010) and Dierwechter and Wessells (2013), the materialized geography of global climate action is the opposite of a donut: the metaphoric hole is the *periphery*, even as larger cities over the past several decades have really become complex city-regions and eco-agglomerative economies. This kind of work on urban(ized) climate action networks, on new efforts by cities (if not yet most city-regions) to address global environmental problems, actually tries to shift the discussion away from New York or Los Angles or Seattle per se, asking wider territorial questions about *city-regional* action similar to those raised earlier by Barrutia et al. (2007) in their important investigations of LA21 in the Basque Country.

This is more than saying that only parts of the globalizing city-regional complex—usually primate or leading cities—are increasingly engaged in forging, joining, and advancing transnational networks of green internationalism. That is an important point on its own, especially in societies where most people live in suburbs and other more peripheral settlements rather than in core cities. As important as New York *City* and its mayors might be, New York is demographically only about a third of the combined statistical area that defines the wider metropolitan region of more than 23 million. It is also saying that we need to pay more attention to how the localized patterns and geographies of putatively green cities shape their international activities, branding, and reputations.

Freiburg, Germany is frequently extolled as Europe's greenest city, which makes it one of the greenest cities in the world, whatever that contested concept may mean to different practitioners and theorists. Yet as Samuel Mossner and Byron Miller (2015) point out, Freiburg's successful experimentations with sustainable urban development from the 1990s "has priced a substantial portion of the population out of the Freiburg housing market" (p. 19). Paradoxically, this has led to a reduction in *urban* social mix within Freiburg as well as greater reliance on "distant municipalities to provide much of the region's affordable housing" which in turn has led to "more extensive commuting, thereby increasing energy consumption and GHG emissions…." (ibid.). Building on their work, we should consider specific cities, even superconductors, within their wider metropolitan or city-regional setting, thinking relationally or even dialectically about how globally engaged cities are locally constituted and non-locally shaped by wider socio-spatial formations.

Like Freiburg, the city of Seattle is often considered one of the most sustainable places in the USA (Portney, 2003). But what *sustains* Seattle's *sustainability* and its active networking and re-positioning in the trendy new archipelagos of green internationalism? Like Freiburg but probably even more so, Seattle—the home of Microsoft and Amazon—has also undergone a profound process of green economic and social transformation over the past few decades. Historically a progressive city, Seattle is nonetheless a paradigmatic case study of how sustainability becomes wrapped into a wider livability machine that attracts high-tech migrants, who then help to put in motion a complex new regionalization of social, economic, eco-technical, and political space (Fig. 4.3). As Dierwechter has previously argued, deploying the work of Allen Scott and others (Dierwechter, 2017, p. 87):

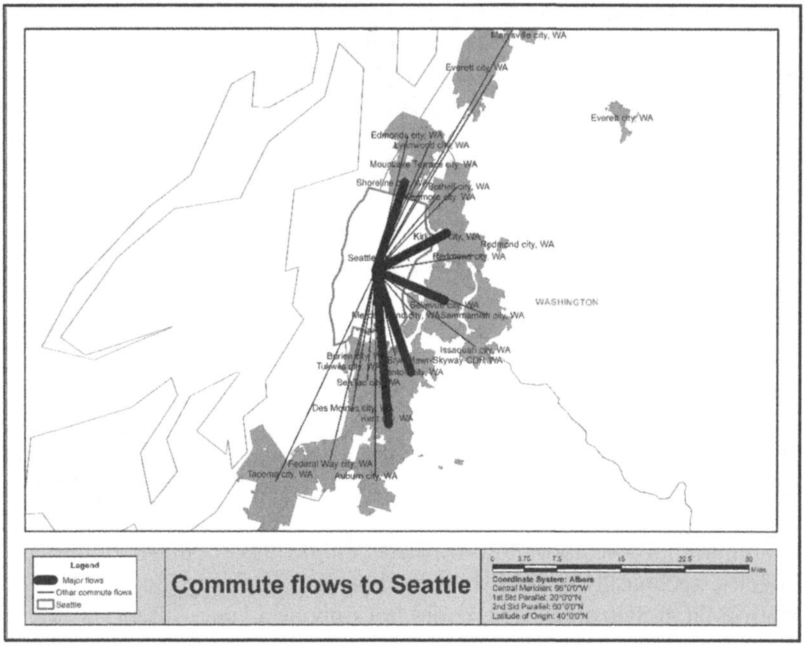

Fig. 4.3 Regional commute flows into Seattle: top 25 places in 2015

> …the regionalization of Seattle's hub-and-spoke economy [inherited from Boeing] over the last few decades represents nothing less than a "world in emergence." 'Seattle' no longer really means only the city, in other words, it means the global city-region, constituted as and by an expansive four-county area. […] In particular, … the 'digitization' of the global economy, the erosion of standardized (Fordist) labor, and the rise of a new class of 'cognitive-cultural' workers like those that populate Microsoft and Amazon are effecting interurban structures of production and work, and, concomitantly, core patterns of urban and regional growth. This is also producing new forms of social stratification (p. 849). Scott [also] argues that a now heavily regionalized and digitized urban social space is being 'radically altered by gentrification, i.e., the colonization of former blue collar neighborhoods in inner-city neighborhoods by members of the cognitariat.' (p. 855)

The patterns of uneven development and the relationships of power between the "cores" of Freiburg and Seattle and parts of their respective "peripheries" provide a kind of metropolitan analog of world systems theory, and perhaps even a structural component of that wider system.

Case study: Transnational Los Angeles in "State-like" California. New York, Freiburg, and Seattle each illuminate important themes in how to engage with the new transnational urban geographies of green internationalism. This section considers the recent experiences of (post) metropolitan or global city-regional Los Angeles, particularly as embedded within California in the wider American federal context. Other than New York, Los Angeles is the most important city in the USA, particularly when conceptually and functionally expanded to include what Edward Soja (2000) and others have theorized of as its "post-metropolitan" spatialities. According to the US Census Bureau, the population of the Metropolitan Statistical Area in 2016, which includes Los Angeles County, was 13,350,000. When expanded to include Riverside-San Bernadino and adjacent urban areas that constitute the wider Combined Statistical Area (CSA), the city-region is much closer to 18 million people. Los Angeles County *alone*, as illustrated in Fig. 4.4, is demographically larger than all but seven other US states—a nation of 327 million people and a $20 trillion economy. (Post-Metropolitan) Los Angeles, technically defined here as Los Angeles-Long Beach-Anaheim Metropolitan Statistical Area, could not only be a large American state in itself; it really could be an entire country had history turned out differently. Indeed, it is now starting to act like one.

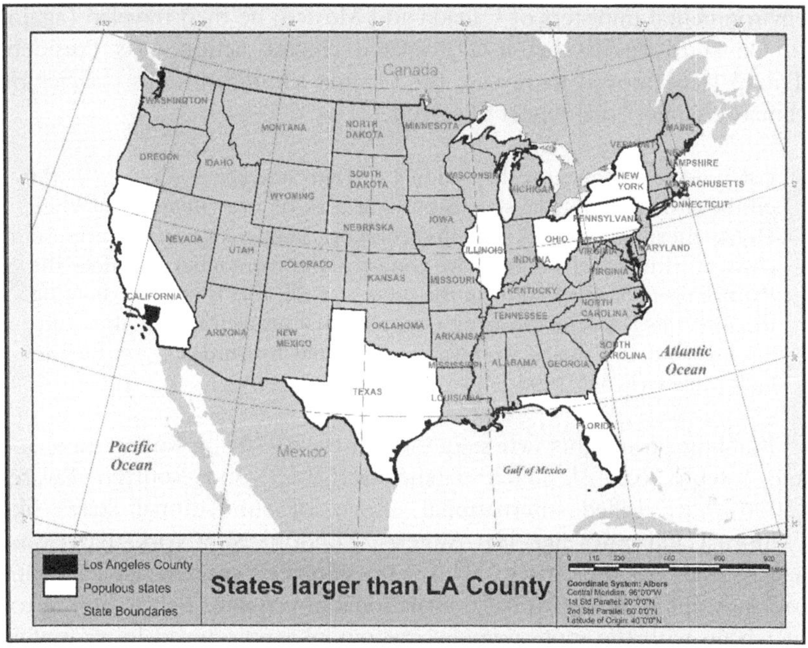

Fig. 4.4 Demographic importance of Los Angeles County in national context

According to Reuters (2016), the state of California, producing about $2.5 trillion in gross state product in 2015, (just) passed the actual nation-state of France in 2016 to become the sixth biggest economy in the world. With a current population of almost 40 million, the politics and policies of the state of California, as earlier stated, have helped to forge what Bansard et al. (2017) consider a massive "superconductor," at least when measured by the density of participation in transnational climate action networks.

Like New York State, but perhaps with even greater political impact, California has also recently thrust a "defiant chin" upwards to President Trump (op. cit.), not least in areas involving global sustainability politics and green policy support. Within months of Trump's election, California had become a "global force" in fighting the Trump administration specifically on climate policy (Davenport & Nagourney, 2017). In May 2017, the governor of California met in San Francisco with the national environmental ministers of Canada and Mexico; he then traveled (again) to China to discuss better-coordinated climate action. "As President Trump [has] moved to reverse the Obama administration's policies on climate change," one report observed,

> California is emerging as the nation's de facto negotiator with the world on the environment. The state is pushing back on everything from White House efforts to roll back pollution rules on tailpipes and smokestacks, to plans to withdraw or weaken the United States' commitments under the Paris climate change accord. In the process, California is not only fighting to protect its legacy of sweeping environmental protection, but also holding itself out as a model to other states — and to nations — on how to fight climate change. (ibid.)

Building on insights first suggested in the 1970s by Robert Keohane and Joseph Nye, Alexander Kuznetsov (2015) and Rodrigo Tavares (2016) have called international efforts of sub-national states like California and cities like Los Angeles, London, New York, Barcelona, and so on, "paradiplomacy." This term signifies a more complex political landscape of international-transnational governance that is (trying to) match up with the more complex cartographies produced by economic and cultural globalization. California's de facto emergence as America's green "negotiator with the world" (op. cit.) is reflected not only its bilateral "state-to-nation" diplomacy, but also in its multilateral efforts to mobilize various kinds of actors associated with cities, the United Nations, major corporations and, at least in theory, congeries of activists who are otherwise also "singers, musicians, mathematicians, professors, students" (Brown, 2017).

The relatively new and increasingly assertive green paradiplomacy of "state-like" polities such as California (Kuznetsov, 2015), while interesting in its own right, is nonetheless relevant to our themes here. As Keating notes, regions like California are "…complex entities containing a multiplicity of groups which may share common interests in some areas but be sharply divided on other issues" (quoted in Kuznetsov, 2015, p. 3). Like New York state, California does not *mainly* govern "up and out" in the diplomatic arena but still "down" and "deep," where differentiated and unevenly urbanized spaces of economic and social life are territorialized. Just as Keating suggests, divisions have emerged in the interstices of these two scalar roles, which also demonstrates theoretically how "the sliding scale of spaces" actively shape the power-geometries and contested politics of "internationalism" in ways we are only just now starting to grasp.

As part of its rising international profile as a "superconductor" of global sustainability and especially carbon mitigation, the California State Legislature has recently considered new legislation—SB [Senate Bill] 827—that would mandate greater density around transit stops (through upzoning, retrofitting, and other local regulatory changes often associated with smart growth) in order to mitigate automobile-based emissions from commuting. While the politics of scale over the (greener) management of urbanization are often diffused and conceptually abstract, consider the following pragmatic paragraph taken directly from the SB 827, as amended through "~~deletions~~ and *additions*" in January 2018 by its co-sponsors:

> This bill would ~~authorize a~~ *require a local government to, if requested, grant a development proponent of a* transit-rich housing project ~~to receive~~ a transit-rich housing ~~bonus~~. *bonus if that development meets specified planning standards, including complying with demolition permit requirements, local inclusionary housing ordinance requirements, preparing a relocation benefits and assistance plan, any locally adopted objective zoning standards, and any locally adopted minimum unit mix requirements.* The bill would define a transit-rich housing project as a residential development project the parcels of which are all within a 1/2 mile radius of a major transit stop or a 1/4 mile radius of a *stop on a* high-quality transit ~~corridor, as those terms are further defined~~. *Corridor* … (see: https://leginfo.legislature.ca.gov/faces/billTextClient.xhtml?bill_id=201720180SB827)

The proposed power shift from "allowing" to "requiring" is significant. Also important is how the state envisions the geometry of sustainable urbanism: nodes and corridors are flanked by housing in precise zones. The multi-scalar pursuit of "urban sustainability through smart growth" has been a signature leitmotif in American urban development practice for a generation (Dierwechter, 2017); but the proposed legislation exposes tensions between global goals like carbon reductions—and "state-like" California as an international green player—and a host of urban-scale prerogatives, both regressive and progressive in nature.

Regressive prerogatives around the locally controlled management of carbon-intensive growth include what Frieden (1979) originally identified across California as the use of local zoning powers to defend privilege, especially with respect to resisting affordable housing. Yet environmental groups like the Sierra Club worry that preemptive infill and developer-oriented density awards "from above" enervate the relevancy of local democracy, while ironically risking new waves of class and race extrusion from extant nodes and corridors through unintentional forms of displacement, gentrification, and elitism (Dougherty & Plumber, 2018). Although SB 827 requires mandatory density bonuses for developers only if "local inclusionary housing" requirements are met, the California chapter of the Sierra Club has opposed this sort of planning and zoning legislation. They argue that high-transit corridors include fixed rail but also bus routes that can "disappear when ridership declines," requiring more flexible, carbon-reducing route adjustments "best left to local community planners" (Jones, 2018). "Regions" and "cities" are both entering the novel world of greener paradiplomacy (Tavares, 2016). But they may not be entering into that diplomacy for the same reasons, on the same "team," with the same sustainability visions, nor the same policy responsibilities, information bases, and political pressures. The urbanization of green internationalism is, once again, subject to the sliding scale of spaces.

Emphasizing "flatter" networks *between* cities co-focused on transnational problems does not change this fundamental proposition (Bouteligier, 2013). Narrowing our discussion again to Metropolitan Los Angeles reveals the importance of the City of Los Angeles as a global node in transnational municipal networks and the uneven intra-metropolitan geographies of green internationalism that rarely get sufficient attention in either the urban or international literatures on environmental governance (and which relate closely to state-based public policies).

The municipality of Los Angeles is one of only twelve large American cities that belong as a US delegation to the C40 Cities transnational municipal network. The other cities are San Francisco, Portland, Seattle, Houston, New Orleans, Chicago, Washington, DC, Philadelphia, New York City, and Boston. Compared with members of the C40 network across the world, the population of Los Angeles, at just over 4 million, is an average-sized participant. Compared with US cities other than New

York and Chicago, however, it is larger; compared especially with other West Coast US cities in the C40 network, the city of Los Angles is far bigger than San Francisco (860,000), Portland (640,000), and Seattle (704,000) *combined*. In this simple demographic sense, climate progress just in the relatively contained mayoral jurisdiction of Los Angeles could potentially represent a substantial American contribution to global action.

Taken on its own terms, the C40 climate network has developed a heuristic tool to monitor progress, assessing members according to key stages of carbon action associated with their Compact of Mayors agreement; these stages are commitment, inventory, target, and plan. They are examples of how cities "act in concert" in global space, self-assessing their own agenda with broadly shared criteria (Davidson & Gleeson, 2015). According to the C40 website, the Compact of Mayors represents "...a common platform to capture the impact of cities' collective actions through standardized measurement of emissions and climate risk, and consistent, public reporting of their efforts, [providing in our view] hard evidence that cities are true climate leaders, and that local action can have a significant global impact."[4] Los Angles was slow to get beyond the easy and cost-free "commitment" stage, initially lagging behind San Francisco, Portland, and Seattle, for instance, all of whom had progressed quickly through the "inventory," "target," and "plan" stages by 2016 (Fig. 4.5). In fairness, giants usually lumber along more slowly.

Stepping outside of these terms of self-assessment, however, various scholars of the C40 experience have been more critical and less celebratory of this "hard evidence" of progress, even as the emergence of this kind of green internationalism is widely considered a key development in the politics, policies, and projects of a new kind of global environmental governance (Chan, 2016; de Macedo, Setzer, & Rei, 2016; Lee, 2013). For Bulkley et al., the C40 Cities Climate Leadership Group represents the global development of a new kind of "strategic urbanism" that differs from earlier forms of "voluntary municipalism" (cited in Davidson & Gleeson, 2015, p. 25). But what does this mean? At the urban scale across the uneven space-economy of Metropolitan Los Angeles, a

[4]This language and related discussion can be found at: http://www.c40.org/programmes/compact-of-mayors.

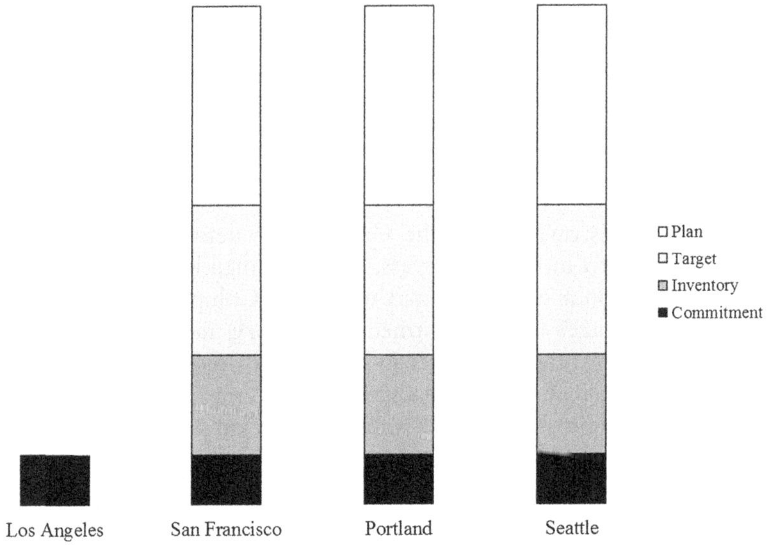

Fig. 4.5 Progress beyond "commitment" to the Compact of Mayors, US West Coast C40 Cities

complex and still poorly understood policy "palimpsest" of green internationalism has emerged, as suggested in Fig. 4.6.

While the big core (or "principal") city of Los Angeles of four million people has emerged as a "strategic" American "node" in the (rising) transnational urbanism of C40 geopolitics, highly uneven efforts to enroll smaller suburban municipalities and larger edge cities in the Mayors Climate Protection Agreement (MCPA) are evident. The MCPA was and remains a US mayoral-led climate action movement initiated by the former mayor of Seattle, Greg Nichols, in (geopolitical) reaction to the George W. Bush administration's infamous refusal to sign the original Kyoto Agreement. Although diverse types of municipalities like Glendora, Pasadena, Long Beach, and Santa Ana have voluntarily signed on to the MCPA, more striking is what might be called the "local divide" in global green responsibilities: i.e., most suburban authorities seem largely indifferent or, much more charitably and likely appropriate, financially/administratively ill-equipped to join the strategic urbanism of C40 Los Angeles (Dierwechter, 2010).

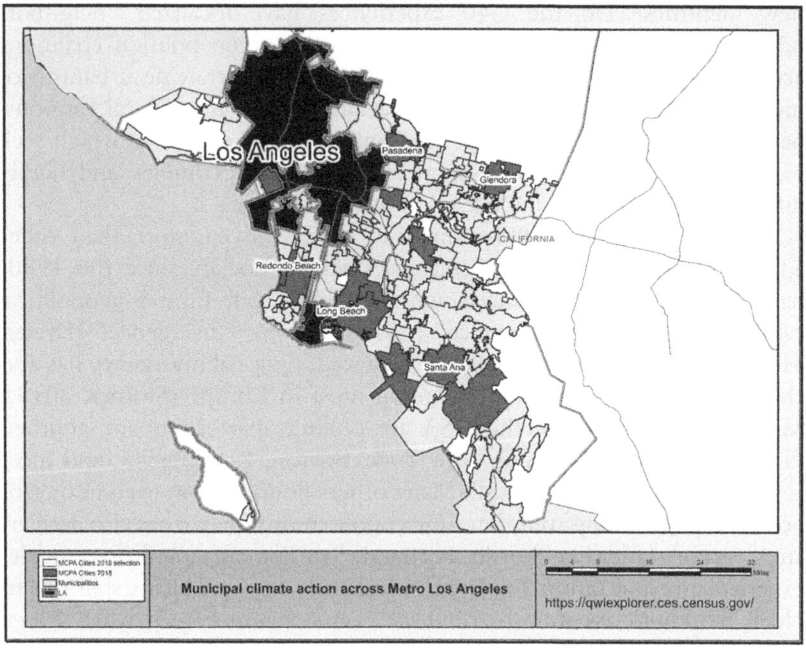

Fig. 4.6 Urban geographies of "strategic" and "voluntary" global climate action across Metropolitan Los Angeles: C40 vs. Mayors Climate Protection Agreement in 2018 (*Source of MCPA data* http://www.mayors.org/climateprotection/ClimateChange.asp)

The greening of cities-regions is generating still other questions. Writing critically from an Australian perspective with specific reference to Sydney, for instance, Davidson and Gleeson (2015) argue that a set of larger structural conditions attending neoliberal capitalism, notably the decisive shift to markets "as a means of governance [,]... has favored the emergence of transnational governance" (p. 23). But does the urbanization of this type of green internationalism in places like Los Angeles (or Sydney) favor greater policy and political fragmentation and unevenness across otherwise integrated city-regional space-economies? Or does it, as with the LA21 experience in the Basque Country discussed earlier (Barrutia et al., 2007), potentially favor new forms of inchoate global city-regionalism and collaborative problem-solving around new identities? Do the C40 experiences have localized "neighborhood effects" on policy action, forging novel green political coalitions across wider metropolitan territories? Or will the strategic urbanism of mega-cities like Los Angeles paradoxically lead to more local tensions, particularly if amplified by state-level initiatives (like SB827) which seek to localize their own power-geometries across the complex and highly differentiated Californian space-economy as a whole?

At lease since the 1980s American society, Even more than other advanced industrialized democracies, American society since the 1980s has experienced a pronounced and disturbing shift from a geopolitical economy favoring "convergence" to one favoring "divergence" (Dreier, Mollenkopf, & Swanstrom, 2014). That said, regional divergence has also characterized other societies, including most in Europe (Storper, 2018). Metropolitan regions in the USA are pulling apart from one another. Places like Seattle, Portland, New York, Boston, Los Angeles, and most C40 cities doing relatively well. Many others flounder. But even in metropolitan regions doing well, intra-metropolitan divergence has also accelerated. As core cities like Seattle, Portland, San Francisco, and Los Angeles experience revitalization—or urban inversion—new problems spill out to older, working-class suburbs (ibid.). A key concern here is what several scholars call the suburbanization of American poverty (Kneebone & Garr, 2010). Storper (2018) sees in the USA and elsewhere a new "density gradient" of employment, income, and opportunity that is now generating and exposing "sharp cleavages in both perceptions and politics" (p. 248).

By and large, these empirical problems have not yet figured prominently in work on urban climate action much less on geographical and political readings of the production and diffusion of new global networks of "green" localities, whether these are focused climate action, like C40, or ancillary social, policy, legal, and political challenges related to urban and national development. Are core cities like C40 cities engaged increasingly with global concerns pulling up and out from their "metropolitan" moorings? If intra-metropolitan *divergence* continues to *grow*, as seems rather likely, how can green policy agendas associated with the global spaces of C40 widen the benefits of participation to poorer, struggling municipalities—some larger than famous cities like Freiburg? Figure 4.7 reimagines the representational geography of metropolitan climate action as a weighted policy landscape of "Yes" and "No," as a metropolitan space-economy in which most local authorities likely do not engage in global issues like climate action even as the core megacity

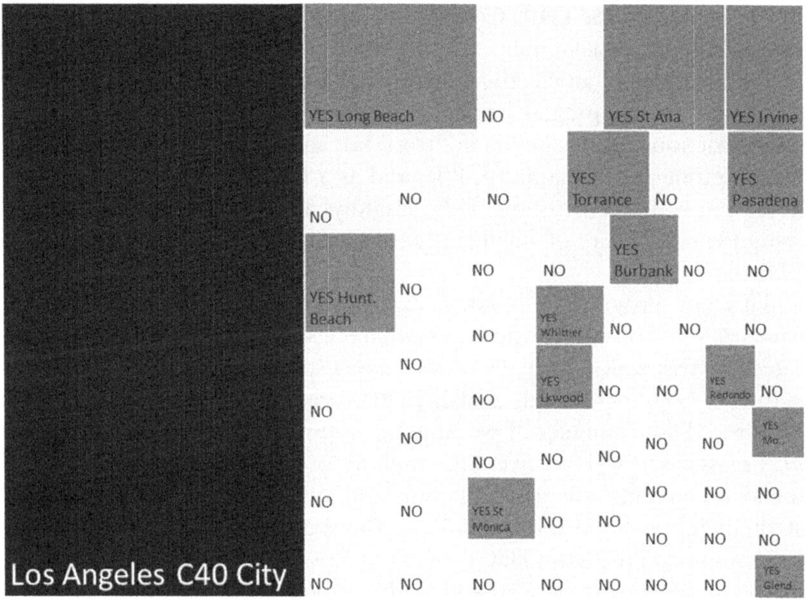

Fig. 4.7 Representational sizes of cities (50,000 or more) in the Los Angeles-Long Beach-Anaheim Statistical Area that had signed on to the Mayors Climate Protection Agreement by 2018 ("yes") vs. those who had not signed the MCPA by 2018 ("no") (*Source of MCPA data* http://www.mayors.org/climateprotection/ClimateChange.asp)

of Los Angeles—the big force below—occupies an increasingly important place in a legitimately new kind of global political and policy action.

As a C40 city, Los Angeles has instituted multiple programs targeting its well-known reputation for over-automobility, congested highways, and recurrent ambient air pollution. *Urban* climate action within Los Angeles, as elsewhere, involves complex policy linkages between global climate change, transportation choices, public health, and uneven social vulnerabilities across the city's neighborhoods. As part of its overall *Sustainable City pLAn* (City of Los Angeles, 2016), ironically called "a roadmap for Los Angles," the city has: created a new Sustainability Officer for the first time in its municipal history; committed to studying "how to achieve the Paris Climate Agreement goal of keeping global temperature rise to 1.5°C" (p. 5); published a greenhouse gas inventory in the service of the C40/Compact of Mayors Global Protocol (p. 35); prioritized the procurement of pure battery electric vehicles (EV) and the deployment of an electric vehicle carshare program (p. 15); designed so-called "environmental justice ordinances" that impose special land-use restrictions on businesses in "toxic hot spots" such as Boyle Heights, Wilmington, and Pacoima (p. 80); and as a final illustration, piloted the development of environmentally friendly affordable housing projects using the city's share of California's substantial "cap-and-trade" funds (p. 77).

This last program is a good example of how a C40-focused Los Angeles—or a C40 city most anywhere else for that matter—complicates our theoretical analyses of the *urban* spaces that *transnational* city networks now increasingly make, particularly as modified by state institutions. "Disadvantaged Los Angeles communities," *Sustainable City pLAn* asserts, "will receive $35 million in cap-and-trade revenues to spend on greenhouse gas reduction and other climate-related projects at the neighborhood level, thanks to the state's Transformative Climate Communities Program (TCC)" (City of Los Angeles, 2016, p. 81). The market-based nature of "cap-and-trade" strategies of carbon governance, rather than, say, a more neo-Keynesian "carbon tax" approach, arguably *envelops* many if not most C40 projects in LA in the geopolitical economies of the global neoliberal project (Davidson & Gleeson, 2015). The city's cap-and-trade revenues, derived from California's emission trading system, reflect a strong marketization of climate policy at all scales (Andrew, Kaidonis, & Andrew, 2010).

On the other hand, the "Transformative Climate Communities Program," like other state-organized initiatives, illustrates how markets are always *shaped* by legislative initiatives that consider other political goals and pressures, including the needs of disadvantaged communities in cities and suburbs. This program is an institutional outcome of California's Global Warming Solutions Act, passed with bi-partisan support in 2006, and known popularly as Assembly Bill (AB) 32. AB32 uses market mechanisms and classic top-down state regulatory power rather than one or the other. In fact, AB 32 was the first program of its kind and scope within the USA to take "a comprehensive, long-term approach" to climate change (see, e.g., https://www.arb.ca.gov/cc/ab32/ab32.htm). Since this time the state of California has targeted major urban and agricultural reductions in carbon dioxide, methane, nitrous oxide, hydrofluorocarbons, perfluorocarbons, sulfur hexafluoride, and nitrogen trifluoride. In their interesting work comparing the subnational diplomacy of the state of California with the province of Sao Paolo, Anderton and Setzer (2017) argue that AB32 is an example of the nascent kinds of "collective entrepreneurship" that increasingly characterize what I am instead calling in this book the new green internationalism in (post-)Westphalian global politics.

Like the politics associated with LA21, the politics of transnational climate action reflect the world's renewed romancing of the city. Like the politics associated with LA21, moreover, this is a chaperoned romance, mediated discursively and supported financially by still durable state or "state-like" institutions, as in California—a polity of nearly 40 people and an overall economy that dwarfs most nation-states. Yet this is a romance that sits rather uneasily on top of the voluminous literature on the politics, economics, ecologies, and especially geographies of post-Metropolitan Los Angeles (Soja, 2010). For Mike Davis (1999, 2006), in arguably still the most devastating portraits of Los Angeles we have, the "ecology of fear" that structures unsustainable accumulation and social disaffection is anything but romantic. From this radical perspective, the polished dashboards of C40 and mayoral globalism are empirically unconvincing, at best, and dangerously Panglossian, at worst: a dreamy world of fantasies and long-distance love letters, not the actually-existing world of power, class, race, and exploitation. Even more "progressive," less dystopian, framings of Los Angeles suggest the complexities and tensions associated how a new transitional green urbanism reshapes metropolitan space and society.

Smart City Urbanization: Mediating Green Corporatism and Eco-(h)ac(k)tivism

Just as the Westphalian urbanization of a greener (but ailing) United Nation system has arguably given more and more ground to the "transnational urbanism" of post-Westphalian networks exploring "glocal" policy action like C40 climate planning, both forms of green internationalism have in turn fallen increasingly in love with all things "smart." The explosion of scholarly and practical attention paid to sexy smart cities, if not yet to wider "smart [city] regions" (Morandi, Rolando, & Di Vita, 2016), is astonishing (Herrschel & Dierwechter, 2018; Hollands, 2015; Townsend, 2013; Wiig, 2015).

Definitions and visions of actual and potential smart cities abound, which means that no one agrees anymore—if they ever did—on exactly *what* they are, much less who they are *for*, or *what* they might imply for wider concerns with global sustainability, urban competitiveness, social justice, psychological-cultural livability, and everyday life chances. Narrowly conceived, smart cities are largely technical artifacts, places that successfully integrate Internet and community (ICT) infrastructures into urban systems and urban behaviors. From this perspective, a smart city

> uses information and communications technologies (ICTs) to make the city services and monitoring more aware, interactive, and efficient.... [The] [s]martness of a city is [thus] driven and enabled technologically by the emergent Internet of Things (IoT)—a radical evolution of the current Internet into a ubiquitous network of interconnected objects that not only harvests information from the environments (sensing) and interacts with the physical world (actuation/command/control), but also uses existing Internet standards to provide services for information transfer, analytics, and applications. (Jin, Gubbi, Marusic, & Palaniswami, 2014, p. 112)

More widely conceived, smart cities are governance innovations, modes of decision-making, and institutional designs that help to facilitate collectively desired changes in the social-material fabric of urbanized societies (Herrschel & Dierwechter, 2015) . In this sense, smart cities are better thought of as strategic governance approaches (Borsekova & Nijkamp, 2018). For Albino, Berardi, and Dangelico (2015),

> Cities worldwide have started to look for solutions which enable transportation linkages, mixed land uses, and high-quality urban services with long-term positive effects on the economy. For instance, high-quality and more efficient public transport that responds to economic needs and connects labor with employment is considered a key element for city growth. Many of the new approaches related to urban services have been based on helping to create what some call "smart cities". (p. 4)

This passage highlights intellectual connections with earlier discussions of smart growth in the USA and Canada. Smartness is definitely, but not only, about "the application of technologies to cities" (p. 10). Invariably, invocations of "smart" as an add-on adjective—smart cities, smart labor, smart economies, smart people, smart growth, smart administration, and so on—has generated "confusion among urban policy makers" (ibid.).

For Albino et al., the term spans two broad "domains" (cf. Lombardi, Giordano, Farouh, & Yousef, 2012; Neirotti, De Marco, Cagliano, Mangano, & Scorrano, 2014). The first is the "hard" domain—the world constituted by "buildings, energy grids, natural resources, water management, waste management, mobility, and logistics, where ICT can play a decisive role in the functions of the systems" (ibid.). The second domain, in contrast, is "soft," and refers to "education, culture, policy innovations, social inclusion, and government, where the applications of ICT are not usually decisive" (ibid.).

Figure 4.8 represents these diverse concerns in a simple way, suggesting that so-called hard approaches to smarter urban spaces and behaviors start with questions of technology and work from there into concerns with governance; alternatively, soft approaches begin with concerns around governance, even as they ultimately must address possibilities afforded by new technologies. Figure 4.8 further suggests that most

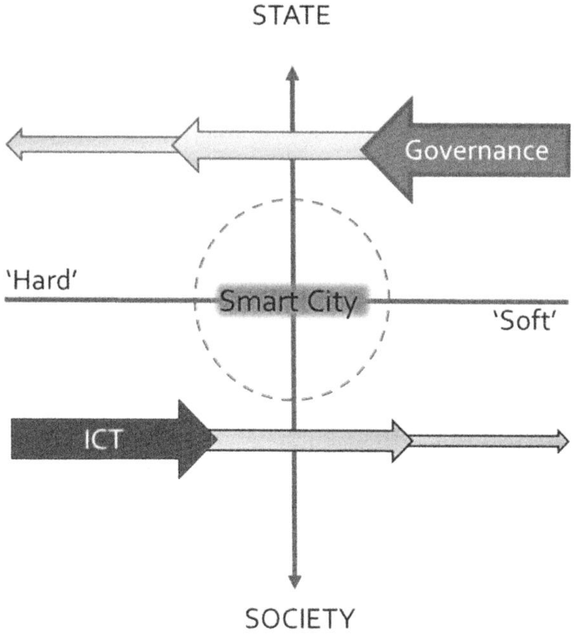

Fig. 4.8 Smartness as urban space: between technology, governance, state, and society

smart city discourses fall somewhere in between the analytical continuum of state and society.

Smart cities, or more accurately smart city urbanization, might therefore emerge more incrementally—service by service; policy arena by policy arena; sector by sector; socio-technical system by socio-technical system—as different actors within and beyond governmental institutions at various scales of operation adopt new technologies and infrastructures of management and democracy. This may or may not be coordinated through comprehensive planning rationalities that seek deliberated synergies and complementarities over time. A great deal of research, particularly in traditionally hard technology fields, has focused on specific services, arenas, sectors, and systems—including work that explores, for instance, the impacts of smart card systems on travel behaviors (Ma, Liu, Wen, Wang, & Wu, 2017) or how "embedded" sensors influence electrical provision in the improved use of buildings (Suryadevara, Mukhopadhyay, Kelly, & Gill, 2015). Much in this area of research and practice arguably relates more accurately to how to make bits and pieces of cities smart*er* through digitization, sensors, actuators, apps, but not necessarily to make smart cities per se.

Alternatively, smart cities as new urban forms (often new towns, although upgrading of in situ spaces is a common trope) might emerge more suddenly and across entire urban systems, at least in theory. Both India and China, who together make up 38% of the world's total population and many of the planet's largest urban areas, are remarkable examples of how the recent obsession with, and politico-economic expectations for, "smart cities" are now components of national urban policies and overall urbanization strategies that simply have so counterpart in, for instance, the USA, the UK, or Australia. Vanessa Watson (2013) has also discussed how African countries—e.g., Kenya, Rwanda, Nigeria—are exploring the rapid development of brand new smart cities as part of their overall management of national economic development through urban planning and human settlements policy. In some cases in all three regions, smart cities merge with parallel discourses of eco-cities, further complicating how the smart turn relates or not to global sustainability (de Jong, Joss, Schraven, Zhan, & Weijnen, 2015).

Under the political leadership of Prime Minister Modi, India launched an enormous urban program called the "100 Smart Cities Mission." Though too soon to evaluate and assess in toto, the mission set out, when launched in 2015, to build 100 new smart cities while renovating/

retrofitting 500 more cities in the coming decades. Various communities from different regions across India have competed with one another for national funding. Each year a list of "winners" emerges. These "winners" are part and parcel of a grand utopian project. In her analysis of planning for India's first smart city—Dholera, in the Ahmedabad district of Gujarat—Ayona Datta (2015, p. 3) traces emerging forms of Indian urbanization though smart cities as a "business model rather than a model of social justice." In her view, the postcolonial Indian state has "internalized the national developmental legacy of utopian urban planning and extended this to a new phase of smart city planning across the country" (p. 17). While the postcolonial Indian state, as Jonas and Moisio (2016) might suggest, is arguably deploying a spatially selective strategy of city-regional development, in general, and smart cities, in particular, as a global geopolitical strategy, Datta specifically shows that, on the one hand, Dholera smart city is "scaling up" a geographically specific "Gujarat model of development" to the nation as a whole (ibid.). On the other hand, drawing on Neil Brenner's work and others, Datta highlights how an

> over-reliance on translocal learning to explain urban change does not shed light on the geographies of land ownership, dispossession, deprivation and struggle generated and entrenched in the unequal distribution of resources and the precarious life conditions against which smart cities like Dholera are conceptualized and materialized. (p. 5)

Marketing for "phase3" of Dholera spatializes how urban forms that ostensibly embody smartness seek to distill and bottle up the "creativity" and glamor of cities from the "caldron" and contradictions of dispossession, deprivation, and struggle, again harkening back to Robert Beauregard's (2018) recent urban intervention into how we romanticize the city. "Dholera smart city phase 3," the developer, SmartHomes infrastructure Pvt. Ltd., aptly promises,

> is an exemplary residential housing township being developed in the first smart city of India: Dholera. The township shall be built using state of art architecture and boast of world class infrastructure; a paradise for those residing within. With ample options to select from, people can purchase varied sizes of plots, smart value apartments, exuberant bungalows and lavish villas. (https://www.dholera-smart-city-phase3.com/)

Dholera phase 3 is a "paradise" because "those residing within" are considered to be a "secured gated society" (ibid.). But urban India is mostly about those "residing without," and the segmentation of smart cities *from* the Indian urban condition raises global questions (Fig. 4.9).

Previous utopian renderings of India's future developments, notably British colonial concerns with imposing a rational urban cartography on a profoundly heterogeneous space and LeCorbusier's post-independence plans for a modernist Chandigarh, in Punjab, provide insights into the challenges that today's smart cities likely face in a globalized world. In his elegiac and brilliant essay on the meaning of the city in India, Sunil Khilnani (1997) juxtaposes the aspirational territorialities of both the colonial/imperial and post-colonial/nationalizing states alongside the actually existing urbanisms of places like Ahmedabad, just to the North of the still-embryonic Dholera smart city. India's cities, Khilnani observes, are "bloated receptacles" for every hope and frustration, housing the

Fig. 4.9 "Residing outside" the smart city: normal urbanization in India (*Source* https://www.pexels.com/public-domain-images/)

"entire compass" of human labor, "from the crudest stone-breaking to the most sophisticated financial transactions" (p. 109). In consequence, while remarkable cities like Ahmedabad were creating their own self-generating modernity through local textile industries—a point not lost on Gandhi—the British kept their distance from the deepest ends of Indian urbanism, while modernizing nationalists like Nehru sought to recast India's great cities into expressions of a proud, post-colonial sovereignty.

Le Corbusier's Chandigarh, built to govern the Punjab province after Lahore was lost to Pakistan in 1947, crystalizes the ongoing risks and realities of today's smart cities, which (once again) Datta notes, have each inherited the urban utopian tradition within India. Chandigarh, Khilnani concludes, was a city "condemned to revolve in an external orbit around the life of its people in Punjab: a glorious stage-set where tableaux of state might be enacted" (p. 130). Nehru chose to build Chandigarh on a site, as he famously proclaimed to the world, "free from existing encumbrances…unfettered by traditions of the past" (cited in Khilnani, 1997, p. 131). "Phase 3" developments in smart cities like Dholera that today market the paradise of "secure gated society" are more historical rhyme than technological future.

In fairness, the larger state(d) aspiration in India is not 100 new smart cities ex nihilo, but 500 in situ experiments in retrofitting and renewing, thus engaging with extant forms of urbanization rather than necessarily condemning places like Dholera to an "external orbit" around the "internal" lives of India's 400 million urban denizens—and counting. So the geographies that smart cities make cannot be collapsed into single cartographic narratives. Certainly building smart cities from scratch—like Howard's garden cities a hundred years ago—suggests the recycled allure and even hubris of totalized city-making, of reshaping flawed pasts into better futures through cleverer presents. Many smart cities so imagined—e.g., Masdar City in Abu Dhabi; Songdo in South Korea; hundreds of new cities in China alone (Neirotti et al., 2014)—run the same risks. The now perfect(ed) future is imaginatively unmoored from past traditions and path-dependent possibilities. But what about the incipient geographies of smart (re)urbanization that are focused instead on the incremental (re)generation of what Araya and Arif (2015) see as the encrusted bureaucratic systems of twentieth-century cities? What do these incipient geographies of smarter renewal—or smartization—reveal about how cities around the world increasingly engage today with global environmental challenges?

For Araya and Arif (2015), smart cities can—and often do—lead to the familiar geographies of "corporatized planning and policy," as many critics of this powerful new trend in international urban development have convincingly hypothesized (Hollands, 2015; Wiig, 2015). However, Araya and Arif also provocatively ask whether—and just how—new forms of smart urbanization might actively facilitate, in certain circumstances, "spaces for democratic practice," where, in their view, "recursive learning" between state and society and what they specifically call "commons-driven government" might flourish going forward (pp. 3–5). These latter themes are worth considering anywhere, especially in the twin giants of India and China, where, along with most of Africa, so much of the globe's ecological and urban future will be made. That said, I now wish to shift the overall discussion to a detailed case study of the smart city experience in Melbourne, Australia, considered by many observers an international exemplar of how urban smartness can help to facilitate global sustainability. In this case, as might be fully expected, tensions between corporatized smartness as a neoliberal form of green politics and the possibilities of a new type of cosmopolitan urban citizenship are revealed. Here, in other words, are contending forms of green internationalism and the city.

Case Study: Melbourne's smart city turn. Like many great cities, Melbourne is like nowhere else but also vaguely, strangely familiar—internationally comparable yet specifically special (Fig. 4.10). Like Seattle, Vancouver, or Cape Town, for instance, Melbourne is an important city within its own national urban system, but not the most important city in that system. Like Seattle, Vancouver, or Cape Town, Melbourne occupies an imagined club of desirable, livable, and trendy cities—hip cities, cool cities—favored by contemporary urban discourses of livability. Like Cape Town, and here rather *unlike* Vancouver, the Canadian city with whom Melbourne is increasingly compared (Holden & Scerri, 2013), Melbourne also has had major, even existential concerns with the sustainability of its regional water supplies, a point raised earlier in this chapter. Under pressure from the so-called Millennium Drought, Melbourne undertook systemic reforms in wastewater recycling, stormwater quality treatment, and stormwater harvesting initiatives (Ferguson, Brown, Frantzeskaki, de Haan, & Deletic, 2013). Like Cape Town, too, Melbourne has a warm temperate climate in the Southern Hemispheric latitudes of the mid-30 degrees, albeit with greater humidity and relatively cooler summers (Cape Town, more like Adelaide, is in the classical Mediterranean climate zone).

Fig. 4.10 Melbourne, Australia: toward a smart green metropolis? (*Source* https://pixabay.com/en/photos/?q=melbourne&hp=&image_type=all&order=&cat=&min_width=&min_height=)

These initial comparisons are apt because families of cities exist within interconnected global urban systems, at least for privileged travelers and elite occupants of specific sectors within global labor markets. These particular cities are neither equatorial nor snow-bound. They share common historical experiences, particularly legal, cultural, and/or linguistic legacies of British colonialism. Such legacies are still more palpable within the old Commonwealth world than in Americanized Seattle. But these are all English-speaking managed cities located along temperate coastlines that emerged from an initial process of land dispossession and cultural alienation from various indigenous peoples (aboriginal, first nation, Native American, San, Khoekhoe). Today, these cities are also all key members of C40—the elite, networked space of urban climate action just discussed. Finally, these are all cities engaged in the smart turn.

Unlike China or India, whose dozens of large cities make up a distributed urban system, Greater Melbourne, at just under 4.5 million people, constitutes nearly one-fifth of the already highly urbanized national population of Australia (the second largest city-region in the USA, Los

Angeles, makes up, even when maximally defined, perhaps 5% of the American national population). In consequence, as a country, Australia's large capital cities—Sydney, Melbourne, Adelaide, Perth, Brisbane, Hobart, Darwin—tend to dominate their respective state populations. The Outback notwithstanding, Australia is a metro nation. Aside from Perth, in Western Australia, Greater Melbourne, in Victoria, has the highest percentage of people in its respective state. About three-quarters of all Victorians live in Greater Melbourne (Fig. 4.11). But do residents of Greater Melbourne live sustainably? Do they live smartly? Are these different questions or just different ways of asking the same thing? What do these questions teach us about the shifting territorialities now redefining cities, states, and global environmental politics?

One way to approach these questions is to suggest that the smart turn across metropolitan Australia, and specifically within Greater Melbourne, is an *extension* of older policy efforts around urban sustainability, including the more recent (but now more than a decade old) internationalism of C40 and other global green policy networking activities. As earlier suggested, the smart turn anywhere can be understood either narrowly (through systemic ICT integration) or more broadly (through

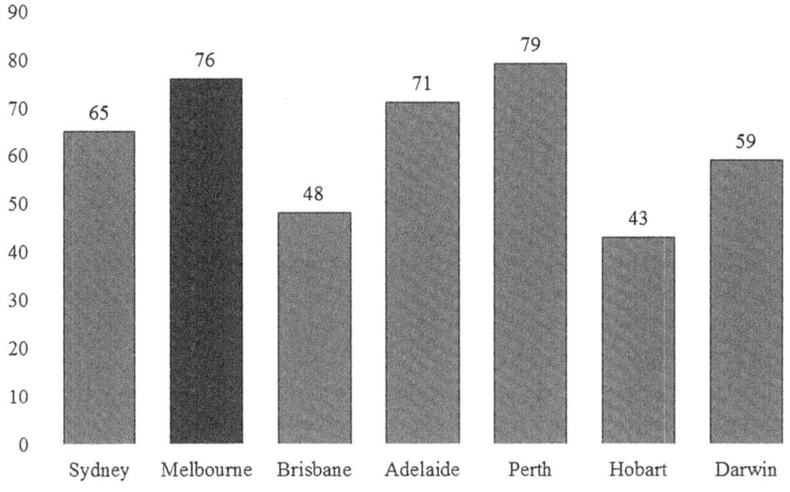

Fig. 4.11 Percentage of overall Australian state populations residing in each capital city (*Source* http://www.abs.gov.au/AUSSTATS/abs@.nsf/mf/3218.0)

governance reforms). However approached, Melbourne is nonetheless frequently singled out—like Vancouver—by the now ubiquitous global rankings of cities on livability scores, sustainability metrics, and similar tropes that often attend parallel concerns with smartness. This overall discourse usefully conditions smartness analyses.

While many global rankings are presented uncritically, skeptics note the deep social costs, economic cleavages, and cultural transitions associated with the highly trendy pursuit of metropolitan livability status. Increasingly priced out of New South Wales and Sydney, migrants to a relatively more livable Melbourne increasingly find a new urban crisis. "Yes Melbourne is liveable – but for whom? Certainly not for everyone," Delaney writes,

> This [was] the year [2017] homeless people became a visible and critical mass in the city. There was no ignoring the tents and tarps ... as the city filled with people paying hundreds of dollars for a seat at the Australian [tennis] Open. It [was] also the year that homeless people were vilified by politicians and the media. That their presence was perceived as some sort of blight. It was the year when property prices in middle-ring suburbs became categorically out of reach of people whose families had lived there for generations. It was when your choice of affordable accommodation became a coffin-like apartment in a city high-rise or a new-build on the city fringe, where you spend your daylight hours at work or in traffic. (Delaney, 2017)

Most of the city-region's working people are in traffic. According to the Australian Bureau of Statistics, 60.2% of people employed aged 15 or over within Greater Melbourne in 2016 drove a car to work, nearly identical to the national average of 61.5%. (As property prices in "middle-ring suburbs" continue to rise, this figure will not diminish much if at all, nor will commute times go down.) As is so often the case, breezy celebrations of livability are both socially and spatially selective—a term of art, many fear, for core-area denizens, as in Freiburg and Seattle, who can afford the great green transition of urban inversion, hop upon usually localized public transit, and enact mixed-use urbanism. Livability is a metropolitan analog of world systems theory, where core-periphery relationships simply play out in a different spatial register.

Melbourne has long engaged in related policy initiatives, such as "knowledge-based urban development" from the 1990s, focusing in part on technology and communication applications to local economies and societies (Yigitcanlar, O'Connor, & Westerman, 2008). In addition, Melbourne has joined global urban policy conversations around learning regions and creative economies (OECD, 2002). But according to Anthopoulos (2017), Melbourne's explicit focus on smart cities emerged only in 2010, about the same time as in London and Seoul but later than Geneva, Tempere, or Hong Kong. In 2010, the city of Melbourne, if not the wider city-region, hosted a C40 workshop. This stimulated new public conversations locally about the potential links between "intelligence" and sustainability. In 2014, Anthopoulos further notes, Melbourne prioritized smart grid deployment and adopted an open data policy. Two years later, the city joined IBM's Smart Cities Challenge. Although focused on the urban core area, the Melbourne City Council's Strategic Planning Office started to define and coordinate a portfolio of smart city projects, such as smart lighting, smart parking, metro Wi-Fi, and smart government services (p. 145). These projects are part and parcel of the transnational urbanism of various global networks focused in some manner on smart policy issues, as already discussed. Melbourne participates in C40 but also United Cities and Local Governments, ISOCARP (a regional planning network), and the Smart Cities Council, among many other global policy networks, conference, and initiatives.

The last of these, the Smart Cities Council, includes an Australia–New Zealand regional network. It provides a very good example of how recent technology discourses around smartness seek to enroll cities into the post-political projects that putatively now constitute how to improve cities (MacLeod, 2013). "We envision a world," the Smart Cities Council proclaims,

> "where digital technology and intelligent design have been harnessed to create smart, sustainable cities with high-quality living and high-quality jobs. To tap into the transformative power of smart technologies, cities need a trusted, neutral advisor. The Smart Cities Council provides that help. We are a network of leading companies advised by top universities, laboratories and standards bodies." (Smart Cities Council, 2016)

Leading companies naturally provide experimenting cities like Melbourne with the mashup allure of corporate-academic-laboratory tested-standardized "readiness guides," "financing templates," "policy frameworks," "visibility campaigns," and "regional networking" (ibid.). This may not quite be *just-add-water-and-stir* governance; not quite

off-the-shelf urbanism—but the overarching themes of salvation through smartness are clear enough. The technical romance with new modalities of urban management expertise is clear enough, too.

Buying and then deploying neutral, post-political products and services concocted elsewhere in a translocated, urbanized world does not mean that (selectively) livable cities like Melbourne *are* post-political spaces. In fact, it *amplifies* the politics at stake. While these products and services help to "reinforce corporatized policy and planning," as once again Araya and Arif (2015) succinctly suggest, cities are geopolitical battlegrounds where open data efforts and many other smart initiatives originating in a local, activist, engaged society of actors can also potentially facilitate new "spaces for citizen empowerment," spaces where the dissemination of smart control systems ultimately "can be subordinated to democratic practices" (op. cit.).

Open data efforts are especially intriguing in this respect.[5] Open data efforts can range from simply publishing government data online (or perhaps hosting code exchanges) to more actively facilitating participatory government. The latter type illustrates how "society-state" relationships can (and sometimes do) change through open data activism (Adams & Adeleke, 2016). In Cape Town, for example, the Social Justice Coalition, a local non-governmental organization, collects data from social media platforms such as Facebook, Twitter, and WhatsApp to communicate how grassroots access to basic sanitation services adversely effects historically disadvantaged communities, especially in former apartheid townships. More generally, NGOs like the Social Justice Coalition "are largely responsible for the open data movement, and the web is now populated by growing numbers of open data catalogs .... [which represents] a tremendously significant development" (Church, 2017, p. 5). The City of Seattle's open data portal, initiated in 2010, symbolizes in a different context the local state's efforts to break down data barriers and information asymmetries. Seattle's open data portal seeks to improve the public's understanding of city operations (and its legal territorialities, e.g., housing code violations by parcel) as well as internal operations that are bureaucratically siloed off from one

[5]This paragraph, particularly references to the Social Justice Coalition, draws directly on multiple conversations and ongoing research collaborations with Britta Ricker of the University of Twente in the Netherlands as well as the author's own research and professional experiences in Cape Town over the past 20 years.

another—now standard digital goals; in addition, it works closely with community-based organizations to bridge the digital divide and advance digital equity (City of Seattle, 2016). Still, the "movement" to bridge state-society chasms and new governance possibilities remain inchoate across the urban world. What Acuto (2018) calls "data-informed urban governance," however populated and deployed, is at present more promise than reality (p. 166).

Cities (and just as likely *actors* embedded within them) are generating, collecting, and sharing more and more data, but smart city-networks, off-the-shelf implementation, new portals, and/or even NGO activism around open data all have major limitations. Interestingly, moving beyond these limitations means circling back to the more traditional forms of internationalism discussed earlier in this chapter no less than current forms of urban transnationalism and different kinds of (smarter) statecraft. "Without effective reform in the UN system and consequent support of national governments," Acuto concludes, "there is little hope for truly global action that goes beyond private interests and networked efforts by cities, which are necessarily selective in the way they connect across borders" (p. 166). Some observers might counter-argue that what is really needed now—given the profundity of the global ecological crisis—is not a different kind of green-internationalized statecraft (albeit through a reformed and more scientifically effective, data-driven United Nations) but, once again, a decentralized, neo-anarchistic confederation of radical municipalities pursuing what Bookchin (1991) originally theorized as an "ecology of freedom."

Less romantic visions, but still challenging to secure in most societies, suggest greater possibilities in a pragmatic series of interrelated governance reforms. Here the smart city is not really a thing—a technical cyber-material fix—but a set of wider changes in how the urban and regional governance of complex places like Greater Melbourne are imagined and conducted. As Herrschel and Dierwechter (2018) argue, smartness is really a way of working not just across technology-society divides, but across spatial and scalar divides that currently occlude the kinds of "transitions" needed to occasion competitiveness, cohesion, and sustainability. Melbourne's *livability* struggles to ensure inclusive prosperity, to win global recognition but also to house its homeless, suggest just how daunting that challenge now is. Shifting the state's digital attention from corporations to citizens, homeless or otherwise, only foregrounds new problems and ethical issues, including the rise of "citizens as sensors," where data are either voluntarily provided or automated in the darkened geographies of digital surveillance (Goodchild, 2007).

Like Cape Town, Melbourne also has numerous NGOs and community groups focused on smart city discourses and social justice concerns of various kinds. Like Seattle and Vancouver, Melbourne's open data portal provides easy dashboard access to digitally active municipal citizens interested in, inter alia, parking dynamics, 3-D building envelopes, bicycling sharing, accessible mobility, sensor feeds, environment and parks, rooftop retrofitting, census changes, land use, and major urban developments. As an example of how technology programs of smartness and wider concerns with global (un)sustainability merge pragmatically in urbanized space, Fig. 4.12 shows a real-time image of Melbourne's Bike Share Stations, with "current number of free and used docks, updated every 15 minutes" (see: https://data.melbourne.vic.gov.au). Clicking on each dot directs users to stations with available bikes. Would that

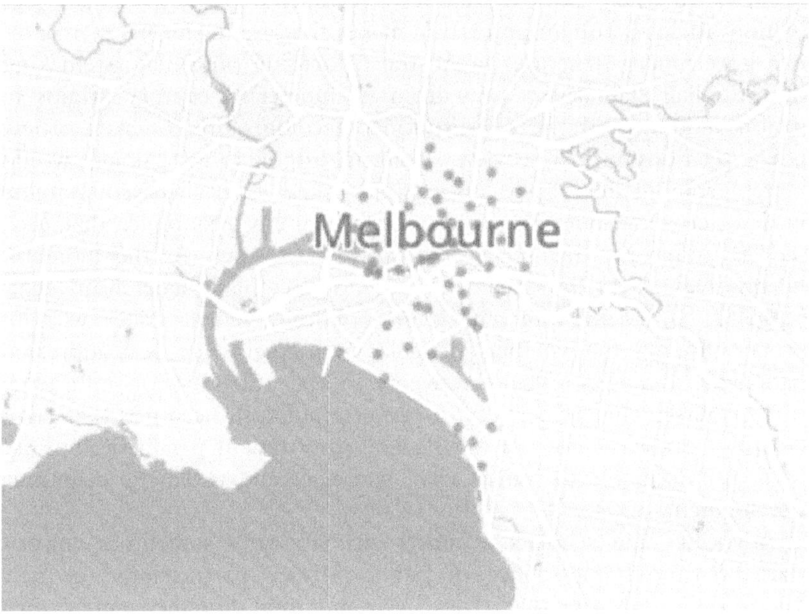

Fig. 4.12 Open data portal products: "real-time" Melbourne Bike Share Stations

Melbourne were only a compact Medieval city of 80,000 rather than polynucleated, globalized city-region of 4.5 million, where most of its suburban workers drive alone rather than bike and where the homeless mostly walk. Both are villainized in the discourses and politics of green internationalism; the former by academicians, the latter by politicians—and both by the median voter.

Interrelated governance reforms in Australia that can amplify the effects of cities as they engage global environmental problems *smartly* therefore imply, at least to some critics, the future development of stronger institutions dedicated to metropolitan governance—or what scholars variously call the New Regionalism, city-regionalism, or just regionalism (Jonas, 2013; Rogerson, 2009; Savitch & Adhikari, 2017; Scott, 2007; Swanstrom, 2006). While places like Greater Melbourne are globalized economic spaces, they are less integrated political or even policy arenas, meaning that smart transitions are generally contained through fragmentation and insufficient subsidiarity. It is the nature of "state space" that is the core problem (Brenner, 1997), in Australia as in the USA, where cities are famously considered creatures of state legislation. Richard Tomlinson (2017) makes the case for formal metropolitan government structures in Australia, focusing on a major remaking of Australian state space away from its nineteenth-century origins in nation-building and the constitutional decolonization of state-society power relationships. In his view, strong metropolitan government would better promote democracy, responsibility, and especially accountability in a society now built largely around capital-city economies that each seek to transform themselves into world-class cities. At the moment, he observes, Australia suffers from "…'opportunistic federalism' characterised by federal intervention in politically salient issues without proper thought for whether particular policy fields are best addressed nationally, locally, or 'cooperatively'…" (p. 150). Major infrastructure disbursements are subject to perverse political incentives that favor "trophy" projects—even as the Federal government paradoxically links such disbursements to "smart city" strategies and neoliberal economic development.

Advocates and scholars of smart city strategies sometimes emphasize the central importance of public–private "partnerships" or new "networks" of actors that cross over and bind together public, private and not-for-profit institutions (Yeh, 2017). Tomlinson's concerns however, rightly situate these arguments within a much wider state-institutional context. Where these "partnerships" and "networks" extend across city-regional space-economies—a big if—they are examples of the so-called New Regionalism that has swept through global governance debates over the past twenty years or so. However, smart partnerships and similar networks in Australia, as Chris McDonald et al. (2008) suggest, are "intermediates" always shaped by the architecture of formal public administrative arrangements.

Conclusions

Actors and institutions at various scales increasingly see urban space as a global solution to ecological challenges. Political ecologies have delocalized and upscaled, a process that has caused its own tensions and political contradictions, made clearer by the shocking populism of 2016, for example, but long-simmering with the impacts of neoliberal globalization (e.g., Dierwechter, 2008; Trapenberg Frick, 2013). This chapter has highlighted the empirical emergence of three (still new) forms or expressions of global green development: *international* urbanization, focusing especially on the urbanization of the United Nations; *transnational* urbanization, focusing on inter-city networks such as C40; and perhaps most recently, *smart city* urbanization, focusing on the generalized infatuation with new Internet and communication technologies of "digitalization" at various territorial scales and modes of governance. Taken together, these new urban expressions reflect less a clean passage from one regime to another than a generalized set of adaptions, transformations, and complex rescalings.

Sustaining human development, though, is not simply a *scalar* debate, not simply a debate about cities versus states vs internationality—about the rise of cities *in and of itself*; it is a debate about new geopolitical economies more generally. Scale alone does not resolve such debates (as if the *polis* is the space of the inherently good, the state the space of the undesirable bad). As argued originally in Chapter 3, Westphalian forms of global politics have hardly vanished from view. State institutions have important effects on post-Westphalian global politics and possibilities—a theme repeatedly demonstrated in this chapter. In consequence, post-Westphalian politics dialectically incorporate and modify Westphalian spatialities; they do not efface them.

References

Acuto, M. (2018). Global science for city policy. *Science (New York, N.Y.), 359*(6372), 165. https://doi.org/10.1126/science.aao2728.

Acuto, M., & Rayner, S. (2016). City networks: Breaking gridlocks or forging (new) lock-ins? *International Affairs, 92*(5), 1147–1166. https://doi.org/10.1111/1468-2346.12700.

Adams, R., & Adeleke, F. (2016). Assessing the potential role of open data in South African environmental management. *The African Journal of Information and Communication, 19*, 79–99.

Albino, V., Berardi, U., & Dangelico, R. M. (2015). Smart cities: Definitions, dimensions, performance, and initiatives. *Journal of Urban Technology, 22*(1), 3–21. https://doi.org/10.1080/10630732.2014.942092.

Anderton, K., & Setzer, J. (2017). Subnational climate entrepreneurship: Innovative climate action in California and São Paulo. *Regional Environmental Change*. https://doi.org/10.1007/s10113-017-1160-2.

Andrew, J., Kaidonis, M., & Andrew, B. (2010). Carbon tax: Challenging neoliberal solutions to climate change. *Critical Perspectives on Accounting, 21*(7), 611–618.

Ansari, S. (2017). The neo-liberal incentive structure and the absence of the developmental state in post-apartheid South Africa. *African Affairs, 116*(463), 206–232. https://doi.org/10.1093/afraf/adw074.

Anthopoulos, L. (2017). Smart utopia VS smart reality: Learning by experience from 10 smart city cases. *Cities, 63*, 128–148. https://doi.org/10.1016/j.cities.2016.10.005.

Araya, D., & Arif, H. (2015). Introduction. In D. Araya (Ed.), *Smart cities as democratic ecologies*. London: Palgrave Macmillan.

Bansard, J., Pattberg, P., & Widerberg, O. (2017). Cities to the rescue? Assessing the performance of transnational municipal networks in global climate governance. *International Environmental Agreements: Politics, Law and Economics, 17*(2), 229–246. https://doi.org/10.1007/s10784-016-9318-9.

Barber, B. (2013). *If mayors ruled the world: Dysfunctional nations, rising cities*. New Haven: Yale University Press.

Barber, B. (2017). *Cool cities: Urban sovereignty and the fix for global warming*. New Haven, CN: Yale University Press.

Barrutia, J., Aguado, I., & Echebarria, C. (2007). Networking for Local Agenda 21 implementation: Learning from experiences with Udaltalde and Udalsarea in the Basque autonomous community. *Geoforum, 38*(1), 33–48. https://doi.org/10.1016/j.geoforuin.2006.05.004.

Barrutia, J., & Echebarria, C. (2011). Explaining and measuring the embrace of Local Agenda 21s by local governments. *Environment and Planning A: Economy and Space, 43*(2), 451–469. https://doi.org/10.1068/a43338.

Barrutia, J., & Echebarria, C. (2013). Why do municipal authorities participate in- and are loyal to- LA21 networks? *Journal of Cleaner Production, 41*, 42–52. https://doi.org/10.1016/j.jclepro.2012.10.006.

Barrutia, J., Echebarria, C., Paredes, M., Hartmann, P., & Apaolaza, V. (2015). From Rio to Rio+ 20: Twenty years of participatory, long term oriented and monitored local planning? *Journal of Cleaner Production, 106*, 594–607. https://doi.org/10.1016/j.jclepro.2014.12.085.

Beauregard, R. (2018). *Cities in the urban age: A dissent*. Chicago: University of Chicago Press.

Bertolini, P., & Giovannetti, E. (2006, July). Industrial districts and internationalization: The case of the agri-food industry in Modena, Italy. *Entrepreneurship & Regional Development, 18*(4), 279–304.

Bond, P. (2010). Maintaining momentum after Copenhagen's collapse: Seal the deal or "Seattle" the deal? *Capitalism Nature Socialism, 21*(1), 14–27. https://doi.org/10.1080/10455751003675839.

Bookchin, M. (1991). *The ecology of freedom: The emergence and dissolution of hierarchy* (Rev ed.). Montreal: Black Rose Books.

Borsekova, K., & Nijkamp, P. (2018). Smart cities: A challenge to research and policy analysis. *Cities, 78*, 1–3.

Bosworth, B. (2016, June 16). South Africa has been key to putting informal settlements on the Habitat III agenda. *Citiscope*. Retrieved from http://citiscope.org/habitatIII/news/2016/06/south-africa-has-been-key-putting-informal-settlements-habitat-iii-agenda.

Bouteligier, S. (2013). Inequality in new global governance arrangements: The north-south divide in transnational municipal networks. *Innovation—The European Journal of Social Science Research, 26*(3), 251–267. https://doi.org/10.1080/13511610.2013.771890.

Brenner, N. (1997). State territorial restructuring and the production of spatial scale: Urban and regional planning in the Federal Republic of Germany, 1960–1990. *Political Geography, 16*(4), 273–306.

Brenner, N. (2014). *Implosions/explosions: Towards a study of planetary urbanization*. Berlin: Jovis.

Brenner, N., & Schmid, C. (2011). Planetary urbanisation. In M. Gandy (Ed.), *Urban constellations*. Berlin: Jovis.

Brown, E. (2017, July 6). *Message to the global citizen festival in Hamburg, Germany*. Retrieved from http://www.latimes.com/politics/essential/la-pol-ca-jerry-brown-hamburg-climate-summit-htmlstory.html.

Bulkeley, H. (2005). Reconfiguring environmental governance: Towards a politics of scales and networks. *Political Geography, 24*(8), 875–902. http://dx.doi.org/10.1016/j.polgeo.2005.07.002.

Busch, H. (2015). Linked for action? An analysis of transnational municipal climate networks in Germany. *International Journal of Urban Sustainable Development*, 1–19. https://doi.org/10.1080/19463138.2015.1057144.

Busch, H. (2018). *Entangled cities: Transnational municipal climate networks and urban governance*. Ph.D. Lund University, Lund, Sweden.

Cape Town City Council. (1998). *The transportation plan for the Wetton-Landsdowne Corridor. Executive summary*. Cape Town: Unpublished Technical Report, City Planner's Department.

Castán Broto, V., & Bulkeley, H. (2013). A survey of urban climate change experiments in 100 cities. *Global Environmental Change, 23*(1), 92–102. http://dx.doi.org/10.1016/j.gloenvcha.2012.07.005.

Chan, D. K. H. (2016). City diplomacy and "glocal" governance: Revitalizing cosmopolitan democracy. *Innovation—The European Journal of Social Science Research, 29*(2), 134–160. https://doi.org/10.1080/13511610.2016.1157684.

Church, J. (2017). Civil society and the open data movement. *DTTP, Documents to the People, 45*(3), 5–8.

City of Cape Town. (2012). *Cape Town spatial development framework: Statutory report*. Cape Town. Retrieved from http://www.capetown.gov.za/work%20and%20business/planning-portal/regulations-and-legislations/cape-town-spatial-development-framework.

City of Cape Town. (2016). *The City of Cape Town's Digital Journey*. Cape Town. Retrieved from http://acceleratecapetown.co.za/wp/wp-content/uploads/2016/11/The-City-of-Cape-Towns-Digital-Journey-Towards-a-Smarter-Future-Rudy-Abrahams-CoCT.pdf.

City of Los Angeles. (2016). *Sustainable city plan: 2nd annual report 2016–2017*. City of Los Angeles. Retrieved from http://plan.lamayor.org/.

City of Seattle. (2016). *Open data program: 2016 annual report*. Seattle: City of Seattle.

Clapp, J., & Dauvergne, P. (2008). *Paths to a green world: The political economy of the global environment*. Cambridge, MA: MIT Press.

Datta, A. (2015). New urban utopias of postcolonial India: 'Entrepreneurial urbanization' in Dholera smart city, Gujarat. *Dialogues in Human Geography, 5*(1), 3–22. https://doi.org/10.1177/2043820614565748.

Davenport, C., & Nagourney, A. (2017, May 23). Fighting Trump on climate, California becomes a global force. *New York Times*. Retrieved from https://www.nytimes.com/2017/05/23/us/california-engages-world-and-fights-washington-on-climate-change.html.

Davidson, K., & Gleeson, B. (2015). Interrogating urban climate leadership: Toward a political ecology of the C40 network. *Global Environmental Politics, 15*(4), 21–38. https://doi.org/10.1162/glep_a_00321.

Davis, M. (1999). *Ecology of fear: Los Angeles and the imagination of disaster*. New York: Vintage Books.

Davis, M. (2006). *City of quartz: Excavating the future in Los Angeles*. London and New York: Verso.

Davis, R. (2013). *Is the future of Cape Town 25km from Cape Town?* Retrieved from https://www.dailymaverick.co.za/article/2013-05-17-is-the-future-of-cape-town-25km-from-cape-town/.

de Jong, M., Joss, S., Schraven, D., Zhan, C. J., & Weijnen, M. (2015). Sustainable-smart-resilient-low carbon-eco-knowledge cities; making sense of a multitude of concepts promoting sustainable urbanization. *Journal of Cleaner Production, 109*, 25–38. https://doi.org/10.1016/j.jclepro.2015.02.004.

de Macedo, L. V., Setzer, J., & Rei, F. (2016). Transnational action fostering climate protection in the city of Sao Paulo and beyond. *Disp, 52*(2), 35–44.

Del Biaggio, C. (2011). Bridging national boundaries: How networks of local actors are building the Alpine region. *Procedia—Social and Behavioral Sciences, 14*, 121–128. https://doi.org/10.1016/j.sbspro.2011.03.029.

Delaney, B. (2017, August 16). Melbourne is 'most liveable city' again. But it's also harder, crueller, out of reach. *The Guardian*. Retrieved from https://amp.theguardian.com/commentisfree/2017/aug/17/melbourne-is-most-liveable-city-again-but-its-also-harder-crueller-out-of-reach.

Dewar, D. (1979). *Low income housing policy in South Africa: With particular reference to the Western Cape/David Dewar, George Ellis*. Cape Town: Urban Problems Research Unit, University of Cape Town.

Dewar, D. (1998, February 8). Personal Communication.

Dewar, D., & Todeschini, F. (1998). *Urban integration and economic development*. London: Frankolin.

Dewar, D., & Watson, V. (1990). *The structure and form of metropolitan Cape Town: Its origins, influences and performance* (Working Paper No. 42). Retrieved from Cape Town.

Dierwechter, Y. (2008). *Urban growth management and its discontents: Promises, practices and geopolitics in US city-regions*. New York: Palgrave.

Dierwechter, Y. (2010). Metropolitan geographies of US climate action: Cities, suburbs and the local divide in global responsibilities. *Journal of Environmental Policy and Planning, 12*(1), 59–82.

Dierwechter, Y. (2017). *Urban sustainability through smart growth: Intercurrence, planning, and the geographies of regional development across Greater Seattle*. Cham, Switzerland: Springer.

Dierwechter, Y., & Wessells, A. (2013). The uneven localisation of climate action in metropolitan Seattle. *Urban Studies, 50*(7), 1368–1385.

Dorogovtsev, S. N., Goltsev, A. V., & Mendes, J. F. F. (2006). K-Core organization of complex networks. *Physical Review Letters, 96*(4), 040601.

Dougherty, C., & Plumber, B. (2018, March 18). A bold, divisive plan to wean Californians from theirs cars. *New York Times*. Retrieved from https://www.nytimes.com/2018/03/16/business/energy-environment/climate-density.html.

Dreier, P., Mollenkopf, J. H., & Swanstrom, T. (2014). *Place matters: Metropolitics for the twenty-first century* (3rd ed.). Lawrence, KS: University Press of Kansas.

Ferguson, B. C., Brown, R. R., Frantzeskaki, N., de Haan, F. J., & Deletic, A. (2013). The enabling institutional context for integrated water management: Lessons from Melbourne. *Water Research, 47*(20), 7300–7314. https://doi.org/10.1016/j.watres.2013.09.045.

Freund, B. (2001). Brown and green in Durban: The evolution of environmental policy in a post-apartheid city. *International Journal of Urban and Regional Research, 25*(4), 717–739.

Frieden, B. (1979). The new regulation comes to suburbia. *The Public Interest, 55*, 15–27.

Frug, G. E. (2008). *City bound: How states stifle urban innovation*. Ithaca: Cornell University Press.

Garcia-Sanchez, I. M., & Prado-Lorenzo, J. M. (2008). Determinant factors in the degree of implementation of Local Agenda 21 in the European Union. *Sustainable Development, 16*(1), 17–34. https://doi.org/10.1002/sd.334.

Gebre-Egziabher, A. (2004). Sustainable cities programme: A joint UN-HABITAT-UNEP facility on the urban environment with participation of the Dutch government. *Annals of the New York Academy of Sciences, 1023*(1), 62–79. https://doi.org/10.1196/annals.1319.016.

Girardet, H. (1999). *Creating sustainable cities*. Foxhole, UK: Green Books Ltd.

Goodchild, M. F. (2007). Citizens as sensors: The world of volunteered geography. *GeoJournal, 69*(4), 211–221. https://doi.org/10.1007/s10708-007-9111-y.

Healey, P. (2006). *Collaborative planning: Shaping places in fragmented societies*. Basingstoke, Hampshire; New York: Palgrave Macmillan.

Herrschel, T., & Dierwechter, Y. (2015). Smart city-regional governance: A dual transition. *Regions, 300*(4), 20–22.

Herrschel, T., & Dierwechter, Y. (2018). *Smart transitions in city-regionalism: The quest for competitiveness and sustainability*. London: Routledge.

Hill, R. C. (2004). Cities and nested hierarchies. *International Social Science Journal, 56*(181), 373–384. https://doi.org/10.1111/j.0020-8701.2004.00500.x.

Holden, M., & Scerri, A. (2013). More than this: Liveable Melbourne meets liveable Vancouver. *Cities, 31*, 444–453. https://doi.org/10.1016/j.cities.2012.07.013.

Hollands, R. G. (2015). Critical interventions into the corporate smart city. *Cambridge Journal of Regions, Economy and Society, 8*(1), 61–77. https://doi.org/10.1093/cjres/rsu011.

Ivanova, M. (2007). Designing the United Nations Environment Programme: A story of compromise and confrontation. *International Environmental Agreements: Politics, Law and Economics, 7*(4), 337–361. https://doi.org/10.1007/s10784-007-9052-4.

Ivanova, M. (2010). UNEP in global environmental governance: Design, leadership, location. *Global Environmental Politics, 10*(1), 30–59. https://doi.org/10.1162/glep.2010.10.1.30.

Jenkins, P., & Wilkinson, P. (2002). Assessing the growing impact of the global economy on urban development in southern African cities—Case studies in Maputo and Cape Town. *Cities, 19*(1), 33–47. https://doi.org/10.1016/s0264-2751(01)00044-0.

Jin, J., Gubbi, J., Marusic, S., & Palaniswami, M. (2014). An information framework for creating a smart city through internet of things. *IEEE Internet of Things Journal, 1*(2), 112–121. https://doi.org/10.1109/jiot.2013.2296516.

Johnson, R. W. (2015). *How long will South Africa survive?* London: Hurst.

Jonas, A. (2013). City-regionalism as a contingent 'geopolitics of capitalism'. *Geopolitics, 18*(2), 284–298. https://doi.org/10.1080/14650045.2012.723290.

Jonas, A. & Moisio, S. (2016). City regionalism as geopolitical processes: A new framework for analysis. *Progress in Human Geography*. https://doi.org/10.1177/0309132516679897.

Jonas, A., While, A., & Gibbs, D. (2004). State modernisation and local strategic selectivity after Local Agenda 21: Evidence from three northern English localities. *Policy & Politics, 32*(2), 151–168. https://doi.org/10.1332/030557304773558125.

Jones, K. (2018). Sierra Club opposes SB 827 [Press release]. Retrieved from http://citizenmarin.org/news/sierra-club-opposes-sb-827/.

Keiner, M., & Kim, A. (2007). Transnational city networks for sustainability. *European Planning Studies, 15*(10), 1369–1395. https://doi.org/10.1080/09654310701550843.

Kern, K., & Bulkeley, H. (2009). Cities, Europeanization and multi-level governance: Governing climate change through transnational municipal networks. *Journal of Common Market Studies, 47*(2), 309–332.

Kern, K., Koll, C., & Schophaus, M. (2007). The diffusion of Local Agenda 21 in Germany: Comparing the German federal states. *Environmental Politics, 16*(4), 604–624. https://doi.org/10.1080/09644010701419139.

Khilnani, S. (1997). *The idea of India*. London: Penguin.

Kneebone, E., & Garr, E. (2010). *The suburbanization of poverty: Trends in metropolitan America, 2000 to 2008*. Washington, DC: Brookings Institution.

Krause, R. M. (2012). An assessment of the impact that participation in local climate networks has on cities' implementation of climate, energy, and transportation policies. *Review of Policy Research, 29*(5), 585–604. https://doi.org/10.1111/j.1541-1338.2012.00582.x.

Krueger, R. (2010). Smart growth and its discontents: An examination of American and European approaches to local and regional sustainable development. *Documents d'Anàlisi Geogràfica, 56*(3), 409–433.

Kuznetsov, A. S. (2015). *Theory and practice of paradiplomacy subnational governments in international affairs*. London and New York: Routledge.

Lancashire County Council. (2003a). *Local Agenda 21—End of term report*. Retrieved from http://www3.lancashire.gov.uk/council/meetings/displayFile.asp?FTYPE=A&FILEID=3329.

Lancashire County Council. (2003b). *Local Agenda 21—Second annual progress report*. Retrieved from www3.lancashire.gov.uk/council/meetings/displayFile.asp?FTYPE=A&FILEID=439.

Le Faye, D. (1995). *Jane Austen's letters*. Oxford: Oxford University Press.

Lee, T. (2013). Global cities and transnational climate change networks. *Global Environmental Politics, 13*(1), 108–127. https://doi.org/10.1162/GLEP_a_00156.

Lee, T., & van de Meene, S. (2012). Who teaches and who learns? Policy learning through the C40 cities climate network. *Policy Sciences, 45*(3), 199–220. https://doi.org/10.1007/s11077-012-9159-5.

Lefebvre, H. (1991). *The production of space*. Oxford, UK: Blackwell.

Lindblom, C. E. (1977). *Politics and markets: The world's political economic systems*. New York: Basic Books.

Llamas-Sanchez, R., Munoz-Fernandez, A., & Maraver-Tarifa, G. (2011). The local Agenda 21 in Andalusia, Spain: A model for sustainable innovation. *African Journal of Business Management, 5*(32), 12653–12663. https://doi.org/10.5897/ajbm11.2381.

Lombardi, P., Giordano, S., Farouh, H., & Yousef, W. (2012). Modelling the smart city performance. *Innovation—The European Journal of Social Science Research, 25*(2), 137–149. https://doi.org/10.1080/13511610.2012.660325.

Ma, X. L., Liu, C. C., Wen, H. M., Wang, Y. P., & Wu, Y. J. (2017). Understanding commuting patterns using transit smart card data. *Journal of Transport Geography, 58*, 135–145. https://doi.org/10.1016/j.jtrangeo.2016.12.001.

MacLeod, G. (2013). New urbanism/smart growth in the Scottish Highlands: Mobile policies and post-politics in local development planning. *Urban Studies, 50*(11), 2196–2221. https://doi.org/10.1177/0042098013491164.

MacLeod, G., & Goodwin, M. (1999). Space, scale and state strategy: Rethinking urban and regional governance. *Progress in Human Geography, 23*(4), 503–527.

Magee, L. (2016). *Interwoven cities*. Houndmills, Basingstoke, Hampshire; New York, NY: Palgrave Macmillan.

Massey, D. (1993). Power-geometries and a progressive sense of place. In J. Bird (Ed.), *Mapping the futures: Local cultures, global change*. London and New York: Routledge.

McDonald, C., Frost, L., Rainnie, A., & Kirk-Brown, A. (2008). *The new regionalism and the role of intermediary organisations in regional development*. Paper presented at the Regional Studies Association Annual Conference, Prague, Czech Republic.

Meadowcroft, J. (2011). Engaging with the politics of sustainability transitions. *Environmental Innovation and Societal Transitions, 1*, 70–75.

Moore, S. (2007). *Alternative routes to the sustainable city: Austin, Curitiba, and Frankfurt*. Lanham, MD: Lexington Books.

Morandi, C., Rolando, A., & Di Vita, S. (2016). *From smart city to smart region: Digital services for an internet of places*. Cham: Springer.

Mossner, S., & Miller, B. (2015). Sustainability in one place? Dilemmas of sustainability governance in the Frieburg Metropolitan region. *Regions, 300*(Winter), 19–21.

Neirotti, P., De Marco, A., Cagliano, A. C., Mangano, G., & Scorrano, F. (2014). Current trends in smart city initiatives: Some stylised facts. *Cities, 38*, 25–36. https://doi.org/10.1016/j.cities.2013.12.010.

Nwaka, G. I. (1996). Planning sustainable cities in Africa. *Canadian Journal of Urban Research, 5*(1), 119–136.

OECD. (2002). *Learning to innovate: Learning regions*. Paris: OECD.

OECD. (2008). *Territorial review: Cape Town metropolitan review*. Paris: OECD.

Ogbazi, J. U. (2013). Alternative planning approaches and the sustainable cities programme in Nigeria. *Habitat International, 40*, 109–118. https://doi.org/10.1016/j.habitatint.2013.03.001.

Osofsky, H. (2005). The geography of climate change litigation: Implications for transnational regulatory governance. *Washington University Law Quarterly, 83*(6), 1789–1856.

Osofsky, H. (2015). Rethinking the geography of local climate action: Multilevel network participation in metropolitan regions. *Utah Law Review, 15*(1), 173–240.

Perlman, B. J., & Jimenez, J. (2010). Creative regionalism: Governance for stressful times. *State and Local Government Review, 42*(2), 151–155. https://doi.org/10.1177/0160323x10380616.

Portney, K. E. (2003). *Taking sustainability seriously: Economic development, the environment, and quality of life in American cities*. Boston: MIT Press.

Pycroft, C. (1998). Integrated development planning or strategic paralysis? Municipal development during the local government transition and beyond. *Development Southern Africa, 15*(2), 151–163.

Rashidi, K., & Patt, A. (2018). Subsistence over symbolism: The role of transnational municipal networks on cities' climate policy innovation and adoption. *Mitigation and Adaptation Strategies for Global Change, 23*(4), 507–523. https://doi.org/10.1007/s11027-017-9747-y.

Reuters. (2016, June 18). California passes France as world's 6th-largest economy. *Rueters*. Retrieved from http://fortune.com/2016/06/17/california-france-6th-largest-economy/.

Roberts, D., & Diederichs, N. (2002). Durban's Local Agenda 21 programme: Tackling sustainable development in a post-apartheid city. *Environment and Urbanization, 14*(1), 189–201.

Rodríguez-Pose, A., & Crescenzi, R. (2008). Mountains in a flat world: Why proximity still matters for the location of economic activity. *Cambridge Journal of Regions, Economy and Society, 1*(3), 371–388. https://doi.org/10.1093/cjres/rsn011.

Rogerson, C. (2004). From spatial development initiative to blue IQ: Subnational economic planning in Gauteng. *Urban Forum, 15*(1), 74–101.

Rogerson, C. (2009). The turn to 'new regionalism': South African reflections. *Urban Forum, 20*(2), 111–140. https://doi.org/10.1007/s12132-009-9057-x.

Rogerson, C., & Rogerson, J. (2015). Johannesburg 2030: The economic contours of a "linking global city". *American Behavioral Scientist, 59*(3), 347–368. https://doi.org/10.1177/0002764214550303.

Savitch, H. V., & Adhikari, S. (2017). Fragmented regionalism. *Journal of Urban Affairs, 53*(2). https://doi.org/10.1177/1078087416630626.

Scott, A. (Ed.). (2001). *Global city-regions: Trends, theory, policy*. Oxford: Oxford University Press.

Scott, J. W. (2007). Smart growth as urban reform: A pragmatic 'recoding' of the new regionalism. *Urban Studies, 44*(1), 15–35.

Scruggs, G. (2017). *How Anne Hidalgo plans to get cities to adopt climate action plans by 2020*. Retrieved from http://citiscope.org/story/2017/how-anne-hidalgo-plans-get-cities-adopt-climate-action-plans-2020.

Setzer, J. (2015). Testing the boundaries of subnational diplomacy: The international climate action of local and regional governments. *Transnational Environmental Law, 4*(02), 319–337. https://doi.org/10.1017/s2047102515000126.

Smart Cities Council. (2016, September 15). About us. *Smart Cities Council*. Retrieved from https://smartcitiescouncil.com/article/about-us-global.

Smith, M. (2001). *Transnational urbanism: Locating globalization*. Malden, MA: Blackwell.

Soja, E. (2000). *Postmetropolis*. London and New York: Sage.

Soja, E. (2010). *Seeking spatial justice*. Minneapolis: University of Minnesota Press.

Sowman, M., & Brown, A. L. (2006). Mainstreaming environmental sustainability into South Africa's integrated development planning process. *Journal of Environmental Planning & Management, 49*(5), 695–712. https://doi.org/10.1080/09640560600849988.

State Capacity Research Project. (2017). *Betrayal of the promise: How South Africa is being stolen*. Retrieved from https://pari.org.za/wp-content/uploads/2017/05/Betrayal-of-the-Promise-25052017.pdf.

Storper, M. (2018). Separate worlds? Explaining the current wave of regional economic polarization. *Journal of Economic Geography, 18*(2), 247–270. https://doi.org/10.1093/jeg/lby011.

Suryadevara, N. K., Mukhopadhyay, S. C., Kelly, S. D. T., & Gill, S. P. S. (2015). WSN-based smart sensors and actuator for power management in intelligent buildings. *IEEE/ASME Transactions on Mechatronics, 20*(2), 564–571. https://doi.org/10.1109/tmech.2014.2301716.

Swanstrom, T. (2006). Regionalism, equality, and democracy. *Urban Affairs Review, 42*(2), 249–257.

Tavares, R. (2016). *Paradiplomacy: Cities and states as global players*. New York: Oxford University Press.

Taylor, P. J. (2012). Extraordinary cities: Early 'City-ness' and the origins of agriculture and states. *International Journal of Urban and Regional Research, 36*(3), 415–447.

Thibert, J. (2015). *Governing urban regions through collaboration: A view from North America*. Burlington, VT: Ashgate.

Toly, N. (2008). Transnational municipal networks in climate politics: From global governance to global politics. *Globalizations, 5*(3), 341–356.

Toly, N. (2017). The new urban agenda and the limits of cities. *The Hedgehog Review, 19*(2), 36–44.

Tomlinson, R. (2017). An argument for metropolitan government in Australia. *Cities, 63*, 149–153. https://doi.org/10.1016/j.cities.2016.10.013.

Törnqvist, G. (2004). Creativity in time and space. *Geografiska Annaler: Series B, Human Geography, 86*(4), 227–243. https://doi.org/10.1111/j.0435-3684.2004.00165.x.

Townsend, A. M. (2013). *Smart cities: Big data, civic hackers, and the quest for a new utopia* (1st ed.). New York: W. W. Norton.

Trapenberg Frick, K. (2013). The actions of discontent: Tea party and property rights activists pushing back against regional planning. *Journal of the American Planning Association, 79*(3), 190–200. https://doi.org/10.1080/01944363.2013.885312.

UN-Habitat. (2012). *History, mandate & role in the UN system*. Retrieved from https://unhabitat.org/history-mandate-role-in-the-un-system/.

United Nations. (2017). *New urban agenda*. Quito: UN Conference of Housing and Sustainable Urban Development. Retrieved from http://habitat3.org/wp-content/uploads/NUA-English.pdf.

Van Hamme, F., & Pion, G. (2012). The relevance of the world-system approach in the era of globalization of economic flows and networks. *Geografiska Annaler: Series B, Human Geography, 94*(1), 65–82.

Van Loon, A. (2018, February 3). Let Cape Town revolutionise the way we think about water. *The Guardian (UK)*. Retrieved from https://www.theguardian.com/commentisfree/2018/feb/04/let-cape-town-revolutionise-the-way-we-think-about-water.

Venter, A. (2007). Integrated development planning as an approach to sustainable development. In G. van Der Waldt (Ed.), *Municipal management: Serving the people* (p. 93). Cape Town: JUTA Press.

Wallerstein, I. M. (1979). *The capitalist world-economy: Essays*. Cambridge, UK: Cambridge University Press.

Wang, V., & McKinley, J. (2018, March 31). In New York budget, a fusillade against de Blasio. *New York Times*. Retrieved from https://www.nytimes.com/2018/03/31/nyregion/new-york-city-budget-cuomo.html.

Watson, V. (1991). Urbanization policy: Lessons from South America for South Africa? *Social Dynamics, 17*(2), 155–167.

Watson, V. (2002). *Change and continuity in spatial planning: Metropolitan planning in Cape Town under political transition*. London and New York: Routledge.

Watson, V. (2013). African urban fantasies: Dreams or nightmares? *Environment and Urbanization, 26*(1), 215–231. https://doi.org/10.1177/0956247813513705.

Weiss, T. G. (2012). *What's wrong with the United Nations and how to fix it* (2nd ed.). Cambridge, UK; Malden, MA: Polity.

While, A., Jonas, A., & Gibbs, D. (2004). The environment and the entrepreneurial city: Searching for the urban 'sustainability; fix' in Manchester and Leeds. *International Journal of Urban and Regional Research, 28*(3), 549–569.

Wiig, A. (2015). The empty rhetoric of the smart city: From digital inclusion to economic promotion in Philadelphia. *Urban Geography*, 1–19. https://doi.org/10.1080/02723638.2015.1065686.

Yeh, A. (2017). *Smart cities in Pacific Asia: Roles of government and private sectors*. Paper presented at the Social and Environmental Implications of Smart Cities: Toward a Global Comparative Research Agenda, University of Calgary.

Yigitcanlar, T., O'Connor, K., & Westerman, C. (2008). The making of knowledge cities: Melbourne's knowledge-based urban development experience. *Cities, 25*(2), 63–72. https://doi.org/10.1016/j.cities.2008.01.001.

Zahran, S., Brody, S. D., Vedlitz, A., Grover, H., & Miller, C. (2008). Vulnerability and capacity: Explaining local commitment to climate-change policy. *Environment and Planning C-Government and Policy, 26*(3), 544–562. https://doi.org/10.1068/c2g.

CHAPTER 5

Conclusions: Global Space and Urban Sustainability

Abstract This final chapter recapitulates the main ideas of the book and assesses the limitations of the arguments on offer. The chapter ends with a brief discussion of a possible research agenda going forward, particularly as this agenda involves a comparative urban studies program increasingly sensitive to the shared geopolitics of environmentalism. Such as program should consider discontent with green policies that are linked discursively to elite actors putatively hostile to industry, nationalism, working-class lives, and "local" democracy, but which arguably now constitute a (once unexpected) part of what Brauch et al. (in: Brauch et al. (eds) Coping with global environmental change, disasters and security, Springer, Berlin, Heidelberg, 2011) have called the "political geoecologies" of the Anthropocene. Here again, and for the final time, the chapter reconsiders how the "abrading and imposition" of different world orders are co-shaping the contemporary interplay between cities, states, and global environmental politics.

Keywords Global sustainability · Rise of cities · Urban geopolitics Stateness

Y. Dierwechter, *The Urbanization of Green Internationalism*, Cities and the Global Politics of the Environment, https://doi.org/10.1007/978-3-030-01015-7_5

A complex, daunting puzzle
Many questions beg for answers—
Sustainability for whom? For how long?
At what rate? [...]
Sustainability, in reality,
Is tantalizing enough, though elusive—
A noble goal, nonetheless...

—C. Gopalakrishnan (2014)

Recapitulations

Poetry nurtures a pathos of optimism to salve the logos of pessimism. Ice sheets the size of entire US states are breaking off Antarctica. Fisheries shrink as deserts grow. Still, city councilors in Portland, Oregon hope that we can "deconstruct" rather than "demolish" older abandoned houses in ways that promote resource reclamation and workforce development within disadvantaged neighborhoods (Balwit, 2017). Global sustainability—one recycled board at a time, one disadvantaged citizen at a time. More generally, Peter Evans (2002, p. 3) has written in his work on urban livability, "it is too soon to dismiss as utopian the possibility of trajectories leading in the direction of greener livability. The possibility of 'green growth machines' or even 'urban livability machines' cannot be ruled out, even in the Third World." Practitioners and academics envision both the theoretical possibilities and empirical pragmatics of urban creativity, but they engage deeply with the injustices and pathologies of the urban cauldron (Beauregard, 2018).

Tensions between passion and logic, between urban creativity and global injustice, pull us in two directions at once. We feel the stress. Writing in the neo-Malthusian tradition, pessimists—particularly natural scientists—tend to emphasize the absolute ecological limits of nature and/or the demographic time bombs of humanity (Foster, 2008; Robertson, 2012). In contrast, optimists, notably economists and technophiles of various kinds, are much more likely to highlight a cornucopia of technological innovation and private goodies—of the arc of our collective history as a bumpy, but still progressive, emancipation from human suffering (Krugman, 2017; Rogers, Jalal, & Boyd, 2005). *In extremis*, global sustainability in certain breezier accounts is a futuristic world of self-driving cars powered by solar-paneled highways or drone-filled airways, where the utopian wonder of science fiction reshapes the urban world of daily life.

In reality, the urbanization of global sustainability is complex. The empirical variety and diffusion of recent developments in world politics increasingly represent, as we head into the second quarter of the century, a legitimate alternative to (if by no means a replacement of) nation-states in the global pursuit of sustainable development goals. This has implications not only for how we talk about the geographies of cities, but also international affairs and especially ecological geopolitics, an important new theme already addressed in different ways by authors such as Jonas (2013), Jonas and Moisio (2016), Herrschel and Newman (2017), and many others (e.g., Barber, 2013; Bouteligier, 2013), including the editors of this series (e.g., Acuto & Rayner, 2016; Setzer, 2015). This multidisciplinary literature, including the books that currently or will soon constitute this book series, represents an exciting new terrain of urban and international scholarship. With the possible exception of the ancient Greek city-states, urbanism has never been more important to wider human-developmental stories.

But this has also transformed how we must approach the politics and geographies of the urban experience. In recent years, controversial concepts like planetary urbanization advance "the radical [Lefebvrean] hypothesis of the complete urbanisation of society, demanding a radical shift in analysis from urban form to the urbanisation process" (Brenner & Schmid, 2011, p. 13). Here cities are no just longer located inside global space: Human global space *itself* is now all but totally urbanized. As a type of human settlement, in other words, urbanization refers to dense agglomerations of interconnected, "interwoven" (Magee, 2016), firms, workers, infrastructures, and otherwise fragmented governments that bind together metropolitan areas and ecological hinterlands, as for instance, in Greater Seattle, Metropolitan Paris, or the European Region of Nuremberg. This is radically far removed from how human beings have lived together for most of human history. While a planet of villages still remains geo-historically legible (Fig. 5.1), a planet of city-regions, of slums and tech hubs, of progressive climate action and regressive resource destruction, is increasingly more accurate—and arguably more alarming.

Still, the anarcho-communitarianism of the village spirit remains a powerful normative ethic and even, for some, a plausible geopolitical imaginary for future green agency. As discussed in Chapter 4, Debbie

Fig. 5.1 Village life in Congo in 1970 (*Credit* Ronald A. Dierwechter; author's collection)

Bookchin (2014) argues *only* for "a global confederation of rebel cities" that rejects *tout court* the bureaucratic discourses of state and international power—whether capitalist, socialist, or other—as "a hopeless pursuit, a dead end." For her and for others, like Ben Barber, the sclerotic state cannot adapt to a world of cities—or to an urbanized society. Perhaps so. Brexit, Trumpism, and nativist–populist politics elsewhere (including even Denmark) give any observer considerable pause; tragically, neo-Fascist states focused on re-bordering a world of flows may not constitute the buried trash heap of history we once imagined them to be. But my abiding hope, as I have argued elsewhere, is that much depends on our current capacity to reimagine and then rebuild over time and across *multiple scales* of regulation and investment,

> a state that is neither overbearing nor invisible; neither exactly neoliberal-competitive nor overtly redistributive or rigidly territorial. Accordingly, a new state space that is: authoritative in knowledge production, but not

> authoritarian in requirement; professional in discourse, but not bureaucratic in disposition; market-respectful, but not market-dominated or wholly "captured"; technocratic in assistance and capable in advise, yet somehow also collaborative, discursive, and resourceful; directive, yet open; pro-growth, yet holistically developmental, protecting resilient ecosystems through improved social cohesion. Here the state is now nimble and relational in support of a theoretically valorized territorial form: the locally competitive yet justly sustainable global city-region. Or so goes the dream [of a smarter state]. (Dierwechter, 2018, p. 54)

The various forms of urbanism, urban development, and urban spaces associated increasingly with the policies and political projects imagined and promoted by international organizations like the United Nations—what I called in Chapter 4 new kinds of international urbanization—demonstrate how the rise of cities is rapidly shifting the priorities and modalities of the traditional interstate system "from above," even as new global spaces of green internationalism—or transnational urbanization—are also produced "from below" by policies and political projects imagined and promoted by networks created by cities themselves. Finally, international, national, and urban actors are explicitly emphasizing smart(er) ICT integration and city-regional governance changes (Herrschel & Dierwechter, 2018). This new global project of "smart city urbanization"—part romance, part romanticization—further integrates planetary dynamics through digital infrastructures without necessarily resolving (and maybe even reproducing) multiple path-dependencies of core-periphery problems; of winners and losers; of local fragmentation and uneven policy and social capacities. As Vanessa Watson (2013) notes in her important study of strategic spatial planning for smart cities across sub-Saharan Africa, it is by no means clear if this putatively green international fantasy is a dream or a nightmare.

At the very least, as I have argued throughout this book, it really is no longer possible to analyze the global politics of the environment without a careful consideration of its various *urbanizations*. Moreover, the emerging urbanizations of global environmental politics are, in my judgment, new empirical reflections of how multiple institutions—including place-based actors, cities, states, and the modern interstate system—are reforming, reassembling, and adapting up, down, and sideways to the multiple threats to economic accumulation, and social cohesion posed by global ecological decay. Finally, the diverse urbanization(s) of a now-shifting green internationalism signals theoretically not a *temporal* succession from a bounded, modern, Westphalian world order to an unbounded, post-Westphalian, networked order; but instead to what Josep Antentas (2015), through Daniel Bensaïd, calls a "sliding scale of spaces."

Bensaid's concept signals the theoretical importance of understanding different temporalities and spatialities of internationalism—including those of the late-adapting state. As suggested in Chapter 2, internationalism usually refers to "the cross-border collective action of subaltern groups" (p. 1102), a definition that could well include the possibility of a confederation of radical municipalities, for instance, as well as disenfranchized groups *emplaced within cities* that act beyond the local scale. Yet internationalism might also refer to "any outlook, or practice, that tends to transcend the nation towards a wider community, *of which nations continue to form the main units*" (p. 1101, emphasis added). We need to attend to all of this.

Anticipations: Researching Green Urban Internationalisms

Further research on the urbanization(s) of green internationalism(s) requires further reflections and critical investigations into the meanings and manifestations of urbanization, and on what constitutes green and especially what (and who) is shaping new types of internationalism—and why, exactly. A relatively short "pivot" is not a long walk, of course, although it might be a helpful step in taking a new direction. I hope so. This Pivot book joins others who see the "rise of cities" as a significant development in world affairs; it sympathizes strongly with recent work in urban political geography that reexamines both the *what* and the *where* of urban politics and policy. In particular, my effort here (to begin) to map out some of the (surely many) new urbanizations of green internationalism largely accepts Ward et al.'s (2018, p. 7) proposition that "... the territorial scope of urban political analysis can no longer be restricted to city limits, especially if those limits are restricted to a matter of formal jurisdictional boundaries." But what, in the end, does this simple and important proposition imply for this book series?

Asking this question does not mean that *places within* cities—e.g., living labs, complete streets, affordable housing projects, slum improvements, informal trading facilities, public squares, clean water pipes, smart parking systems, phones-in-ambulatory hands, and so on—are unimportant; far from it. But the geopolitics and multi-scalar spatialities of the "urban" now clearly stretch well beyond the familiar territorialities of city limits and not simply to the wider agglomeration economies of entire city-regions of different sizes—Freiburg, Seattle, Los Angeles, Cape Town, Melbourne, for instance. Without question, the politics and spatialities of urban life are now reproduced as unevenly constructed transnational networks like C40 that bind such places into global discourses of urbanized sustainability. The "global urban" of these networks—while new and somewhat strange—are nonetheless familiar in very many ways: power-geometries of uneven development, for example, do not melt away because "mayors rule the worlds." Some cities are more connected and more important, even as they are (and maybe *because* they are) territorialized by wider political communities, including central states and "state-like" complexes, such as shown in the discussion of California. What, accordingly, is really new here? Are these networks legitimately innovative spaces of global sustainability and revived democracy or just the latest political iterations required for the accumulation that will eventually destroy us?

Moreover, the rise of many cities in global environmental affairs may well be tied closely to the local reproduction of regional-scale inequalities and path dependencies. What does it take, I asked in Chapter 4, to *sustain sustainability* in, say, Seattle, Freiburg, or Melbourne? "Sliding down" from the well-documented pathologies of contemporary nation-states should never blind us to the various ways in which classical power relations of uneven property ownership—of land—use patterns and standardized forms of local regulatory power, such as exclusionary zoning—obviates socially just transitions in sustainability. Who is empowered by the rise of cities? Who is not? Are these co-constitutive? How do or how can—indeed, *can*—multi-level states mitigate the perhaps unintended consequences associated with the presumed benefits of policy localizations that some feel directly facilitates greater flexibility and program creativity? Relatedly, how can we measure developments in green internationalism, both quantitatively and qualitatively?

Finally, as part of what Brauch et al. (2011) have called the new "political geoecologies" of the Anthropocene, the rise of cities in global environmental politics seems dialectically tied up with what Andres Rodríguez-Pose (2018) documents as the populist revenge of "the places that don't matter," whether understood as Brexit or Trumpism. What I have repeatedly highlighted as the abrading and imposition of different world orders are co-shaping the contemporary interplay between cities, states, and global environmental politics—and these dynamics clearly do not constitute a singular, linear story. How, then, can elites in favored cities avoid charges of arrogance and hostility for working-class lives? How can green policies include blue collars?

These questions and many others are increasingly important as global hopes of sustainability fade, as some argue that it is either "too late for sustainability" (Bullard, 2011) or that, less dismissively, "it is time to move past the concept of sustainability" (Benson & Craig, 2014). But if *not* sustainability—or sustainable urban development or green urbanism—what? exactly? Livability? Resiliency? Post-capitalism? Such concepts *theoretically* might improve upon an admittedly threadbare and attenuated discourse, but the empirics of the ecological problematic, from carbon loading to species extinction, remain stubborn and long-running. They remain *facts*, it must be said, even in a post-factual world of tweets. In this book, human development refers fundamentally (not contingently) to the economic transformation of nature into specific kinds of societies, an *old* process that nonetheless, since around 1500 and the remarkable rise of capitalist (neo)colonization, has unquestionably involved planetary-wide dynamics, or what we now think of as international geographies of economic-ecological articulation and political contestation. Do the re-scaled ***urbanizations*** of internationalism—through the urbanization of the United Nations and its more urbanized member states, for example, or through the transnationalism of networked municipalities "teaching and learning" from one another, or through the recent search for "smarter" forms of living—in turn make us greener?

Complex, daunting, tantalizing, elusive—yet noble questions indeed.

References

Acuto, M., & Rayner, S. (2016). City networks: Breaking gridlocks or forging (new) lock-ins? *International Affairs, 92*(5), 1147–1166. https://doi.org/10.1111/1468-2346.12700.

Antentas, J. (2015). Sliding scale of spaces and dilemmas of internationalism. *Antipode, 47*(5), 1101–1120. https://doi.org/10.1111/anti.12169.

Balwit, N. (2017). *Tearing down old homes, building a workforce.* Retrieved from https://www.citylab.com/life/2017/06/portland-deconstruction-demolition-workforce-training/529902/?utm_source=feed.

Barber, B. (2013). *If mayors ruled the world: Dysfunctional nations, rising cities.* New Haven: Yale University Press.

Beauregard, R. (2018). *Cities in the urban age: A dissent.* Chicago: University of Chicago Press.

Benson, M. H., & Craig, R. K. (2014). The end of sustainability. *Society & Natural Resources, 27*(7), 777–782. https://doi.org/10.1080/08941920.2014.901467.

Bookchin, D. (2014). *Radical municipalism: The future we deserve.* Retrieved from https://www.filmsforaction.org/articles/radical-municipalism-the-future-we-deserve/.

Bouteligier, S. (2013). Inequality in new global governance arrangements: The north-south divide in transnational municipal networks. *Innovation—The European Journal of Social Science Research, 26*(3), 251–267. https://doi.org/10.1080/13511610.2013.771890.

Brauch, H., Dalby, S., & Spring, Ú. (2011). Political geoecology for the anthropocene. In H. G. Brauch, Ú. Oswald Spring, C. Mesjasz, J. Grin, P. Kameri-Mbote, B. Chourou, … J. Birkmann (Eds.), *Coping with global environmental change, disasters and security* (Vol. 5, pp. 1453–1485). Berlin, Heidelberg: Springer.

Brenner, N., & Schmid, C. (2011). Planetary urbanisation. In M. Gandy (Ed.), *Urban constellations.* Berlin: Jovis.

Bullard, N. (2011). It's too late for sustainability. What we need is system change. *Development, 54*(2), 141–142.

Dierwechter, Y. (2017). The smart state as utopian space for urban politics. In A. Jonas, B. Miller, K. Ward, & D. Wilson (Eds.), *The Routledge handbook on spaces of urban politics.* London and New York: Routledge.

Evans, P. (Ed.) (2002). *Livable cities? Urban struggles for livelihood and Sustainability.* Berkeley University of California Press.

Foster, J. (2008). *The sustainability mirage.* London: Earthscan.

Gopalakrishnan, C. (2014). Poem: On sustainability. *Journal of Natural Resources Policy Research, 6*(2–3), 213–213. https://doi.org/10.1080/19390459.2014.924250.

Herrschel, T., & Dierwechter, Y. (2018). *Smart transitions in city-regionalism: The quest for competitiveness and sustainability.* London: Routledge.

Herrschel, T., & Newman, P. (2017). *Cities as international actors: Urban and regional governance beyond the nation state.* London and New York: Palgrave.

Jonas, A. (2013). City-regionalism as a contingent 'geopolitics of capitalism'. *Geopolitics, 18*(2), 284–298. https://doi.org/10.1080/14650045.2012.723290.

Jonas, A. & Moisio, S. (2016). City regionalism as geopolitical processes: A new framework for analysis. *Progress in Human Geography.* https://doi.org/10.1177/0309132516679897.

Krugman, P. (2017, June 1). Trump gratuitously rejects the Paris Climate Accord. *New York Times*. Retrieved from https://www.nytimes.com/2017/06/01/opinion/trump-gratuitously-rejects-the-paris-climate-accord.html?_r=0.

Magee, L. (2016). *Interwoven cities*. Houndmills, Basingstoke, Hampshire; New York, NY: Palgrave Macmillan.

Robertson, T. (2012). *The Malthusian moment: Global population growth and the birth of American environmentalism*. New Brunswick: Rutgers University Press.

Rodríguez-Pose, A. (2018). The revenge of the places that don't matter (and what to do about it). *Cambridge Journal of Regions, Economy and Society, 11*(1), 189–209.

Rogers, R., Jalal, K., & Boyd, J. (2005). *An introduction to sustainable development*. Cambridge, MA: London Harvard University Press.

Setzer, J. (2015). Testing the boundaries of subnational diplomacy: The international climate action of local and regional governments. *Transnational Environmental Law, 4*(02), 319–337. https://doi.org/10.1017/s2047102515000126.

Watson, V. (2013). African urban fantasies: Dreams or nightmares? *Environment and Urbanization, 26*(1), 215–231. https://doi.org/10.1177/0956247813513705.

Ward, K., Jonas, A., Miller, B., & Wilson, D. (2018). The spaces of urban politics: An introduction. In K. Ward, A. Jonas, B. Miller, & D. Wilson (Eds.), *The Routledge handbook on spaces of urban politics*. New York: Routledge.

Glossary of Key Terms

Urbanization i. Traditionally, the long-term historical process whereby once predominately rural populations increasingly become urban populations, although what constitutes "urban" is defined differently in different societies (e.g. at least 10,000 people in Sweden; 5000 in India). The "level" of urbanization in a society refers to the percentage of the total national population that resides in cities. The "rate" of urbanization is expressed as an annual percentage of change.

ii. In this book, the urban expressions, diverse confluences, emplacement, and/or materializations of socio-political relations stretched out across territorialized scales of power and regulation.

iii. Planetary urbanization refers less to the urban as a place than to urbanization as a worldwide process of change that has effectively enveloped human development activities nearly everywhere. In this sense, Iowa is less urban than 'urbanized.' Farmers are "plant managers" in both the agricultural and industrialized sense of the term.

Global ecological politics A relatively new set of actual practices within international relations that focuses on issues related to the interaction of humans and the natural world. Prior to the 1980s, most work related specifically to resource extraction and development issues. Only in the 1980s and 1990s did global environmental/ecological politics emerge as a distinct field looking at truly "global" problems

Y. Dierwechter, *The Urbanization of Green Internationalism*, Cities and the Global Politics of the Environment, https://doi.org/10.1007/978-3-030-01015-7

like ozone depletion, climate change, ocean acidification, deforestation, and desertification

Green internationalism A term specific to this book that is roughly interchangeable with 'global ecological politics,' but that signals a desire to move away from strictly 'Westphalian' modes of analysis and thinking, and that while retaining state actors, links into long-standing forms of trans-local and cross-border activities within and between cities, scientists, trade unions, business groups, and social reformers of all kinds.

Westphalian world order The 'taken-for-granted' world political system that was established very slowly after 1648 as part of the Peace of Westphalia, wherein three main theoretical principles define the world "order": (1) state sovereignty; (2) (legal) equality of states; (3) non-intervention of one state in the affairs of another.

Post-Westphalian world order A new set of processes that are challenging the territorial and political foundations of the Westphalian world order, evident since the 1970s according to some researchers. The most recent transformation of the world order is, in other words, often depicted as a shift from a Westphalian to a post-Westphalian era in which *international organizations (of various kinds)* are becoming increasingly independent sites of authority. Associated with the "rise of cities" thesis.

International urbanization Forms of urbanism, urban development, urban space etc. associated with the policies and political projects imagined and promoted by international organizations like the UN, e.g. urbanism in SDGs.

Transnational urbanization Forms of urbanism, urban development, urban space, etc. associated with the policies and political projects imagined and promoted by networks largely created by cities themselves.

Smart city urbanization Forms of urbanism, urban development, etc. associated with the policies and political projects imagined and promoted by international, national, and urban actors that explicitly emphasize ICT integration and regional governance changes.

Development The economic transformation of nature into economies and societies.

City-regions Dense agglomerations of firms, workers, infrastructures, and otherwise fragmented governments that bind together metropolitan areas and their various ecological hinterlands (e.g. Greater Seattle,

Metropolitan Paris, Cape Town Functional Region, European Region of Nuremberg).

The State A human community, for Max Weber, that claims legitimate monopoly over the means of coercion in a given territorial area. For Bob Jessop, more simply, "the territorialization of political authority" which necessarily involves "the intersection of politically organized coercive and symbolic power, a clearly demarcated core territory, and a fixed population on which political decisions are collectively binding."

Bibliography

Acuto, M. (2018). Global science for city policy. *Science (New York, N.Y.), 359*(6372), 165. https://doi.org/10.1126/science.aao2728.

Acuto, M., & Rayner, S. (2016). City networks: Breaking gridlocks or forging (new) lock-ins? *International Affairs, 92*(5), 1147–1166. https://doi.org/10.1111/1468-2346.12700.

Adams, R., & Adeleke, F. (2016). Assessing the potential role of open data in South African environmental management. *The African Journal of Information and Communication, 19*, 79–99.

Agnew, J. (1994). The territorial trap: The geographical assumptions of international relations theory. *Review of International Political Economy, 1*(1), 53–80. https://doi.org/10.2307/4177090.

Agnew, J. (2000). Global political geography beyond geopolitics. *International Studies Review, 2*(1), 91–99.

Albino, V., Berardi, U., & Dangelico, R. M. (2015). Smart cities: Definitions, dimensions, performance, and initiatives. *Journal of Urban Technology, 22*(1), 3–21. https://doi.org/10.1080/10630732.2014.942092.

Alyanak, L., & Cruz, A. (1997). Implementing Agenda 21: NGO experiences from around the world. *Environment and Urbanization, 9*(2), 322–323.

Amen, M., Toly, N., McCarney, P., & Segbers, K. (Eds.). (2012). *Cities and global governance: New sites for international relations*. Farnham, UK: Ashgate.

Anderson, B. (2016). *Imagined communities : Reflections on the origin and spread of nationalism*. London: Verso.

Y. Dierwechter, *The Urbanization of Green Internationalism*, Cities and the Global Politics of the Environment, https://doi.org/10.1007/978-3-030-01015-7

Anderton, K., & Setzer, J. (2017). Subnational climate entrepreneurship: Innovative climate action in California and São Paulo. *Regional Environmental Change*. https://doi.org/10.1007/s10113-017-1160-2.

Andrew, J., Kaidonis, M., & Andrew, B. (2010). Carbon tax: Challenging neoliberal solutions to climate change. *Critical Perspectives on Accounting, 21*(7), 611–618.

Ansari, S. (2017). The neo-liberal incentive structure and the absence of the developmental state in post-apartheid South Africa. *African Affairs, 116*(463), 206–232. https://doi.org/10.1093/afraf/adw074.

Antentas, J. (2015). Sliding scale of spaces and dilemmas of internationalism. *Antipode, 47*(5), 1101–1120. https://doi.org/10.1111/anti.12169.

Antentas, J. (2016). Daniel Bensaïd, Melancholic strategist. *Historical Materialism, 24*(4), 51–106. https://doi.org/10.1163/1569206x-12341500.

Anthopoulos, L. (2017). Smart utopia VS smart reality: Learning by experience from 10 smart city cases. *Cities, 63*, 128–148. https://doi.org/10.1016/j.cities.2016.10.005.

Appadurai, A. (1996). *Modernity at large cultural dimensions of globalization*. Minneapolis: University of Minnesota Press.

Araya, D., & Arif, H. (2015). Introduction. In D. Araya (Ed.), *Smart cities as democratic ecologies*. London: Palgrave Macmillan.

Attwell, W. (2014). The rise of cities as global actors: What consequences for policy? *Global Policy, 5*(3), 374–376.

Backstrand, K., & Kronsell, A. (2015). *Rethinking the green state: Environmental governance towards climate and sustainability transitions*. London and New York: Routledge.

Bailey, I., Gouldson, A., & Newell, P. (2011). Ecological modernisation and the governance of carbon: A critical analysis. *Antipode, 43*(3), 682–703.

Bairoch, P. (1988). *Cities and economic development: From the dawn of history to the present*. Chicago: University of Chicago Press.

Balwit, N. (2017). *Tearing down old homes, building a workforce*. Retrieved from https://www.citylab.com/life/2017/06/portland-deconstruction-demolition-workforce-training/529902/?utm_source=feed.

Bansard, J., Pattberg, P., & Widerberg, O. (2017). Cities to the rescue? Assessing the performance of transnational municipal networks in global climate governance. *International Environmental Agreements: Politics, Law and Economics, 17*(2), 229–246. https://doi.org/10.1007/s10784-016-9318-9.

Bansard, J. S., Pattberg, P. H., & Widerberg, O. (2017). Cities to the rescue? Assessing the performance of transnational municipal networks in global climate governance. *International Environmental Agreements: Politics, Law and Economics, 17*(2), 229–246. https://doi.org/10.1007/s10784-016-9318-9.

Barash, D. (2001, February 23). Buddhism and the subversive science. *Chronicle of Higher Education*. Retrieved from https://www.chronicle.com/article/Buddhismthe-Subversive/32948.

Barber, B. (2013). *If mayors ruled the world: Dysfunctional nations, rising cities*. New Haven: Yale University Press.

Barber, B. (2017). *Cool cities: Urban sovereignty and the fix for global warming*. New Haven, CN: Yale University Press.

Barchiesi, F. (2001). Transnational capital, urban globalisation and cross-border solidarity: The case of the South African municipal workers. In P. Waterman & J. Wills (Eds.), *Place, space and the new labour internationalisms*. Oxford, UK; Malden, MA: Blackwell.

Barrutia, J., Aguado, I., & Echebarria, C. (2007). Networking for Local Agenda 21 implementation: Learning from experiences with Udaltalde and Udalsarea in the Basque autonomous community. *Geoforum, 38*(1), 33–48. https://doi.org/10.1016/j.geoforuin.2006.05.004.

Barrutia, J., & Echebarria, C. (2011). Explaining and measuring the embrace of Local Agenda 21s by local governments. *Environment and Planning A: Economy and Space, 43*(2), 451–469. https://doi.org/10.1068/a43338.

Barrutia, J., & Echebarria, C. (2013). Why do municipal authorities participate in- and are loyal to- LA21 networks? *Journal of Cleaner Production, 41*, 42–52. https://doi.org/10.1016/j.jclepro.2012.10.006.

Barrutia, J., Echebarria, C., Paredes, M., Hartmann, P., & Apaolaza, V. (2015). From Rio to Rio+ 20: Twenty years of participatory, long term oriented and monitored local planning? *Journal of Cleaner Production, 106*, 594–607. https://doi.org/10.1016/j.jclepro.2014.12.085.

Beatley, T. (1999). *Green urbanism: Learning from European cities*. Washington, DC: Island Press.

Beauregard, R. (2018). *Cities in the urban age: A dissent*. Chicago: University of Chicago Press.

Benson, M. H., & Craig, R. K. (2014). The end of sustainability. *Society & Natural Resources, 27*(7), 777–782. https://doi.org/10.1080/08941920.2014.901467.

Bertolini, P., & Giovannetti, E. (2006, July). Industrial districts and internationalization: The case of the agri-food industry in Modena, Italy. *Entrepreneurship & Regional Development, 18*(4), 279–304.

Bodoni, C. (2013, February 12). *A crisis of the state? The end of the postWestphalian model*. Retrieved from http://www.socialeurope.eu/author/carlo-bordoni/.

Bond, P. (2000). *Elite transition: From apartheid to neoliberalism*. London: Pluto Press.

Bond, P. (2010). Maintaining momentum after Copenhagen's collapse: Seal the deal or "Seattle" the deal? *Capitalism Nature Socialism, 21*(1), 14–27. https://doi.org/10.1080/10455751003675839.

Bookchin, D. (2014). *Radical municipalism: The future we deserve*. Retrieved from https://www.filmsforaction.org/articles/radical-municipalism-the-future-we-deserve/.

Bookchin, M. (1991). *The ecology of freedom: The emergence and dissolution of hierarchy* (Rev ed.). Montreal: Black Rose Books.

Borsekova, K., & Nijkamp, P. (2018). Smart cities: A challenge to research and policy analysis. *Cities, 78*, 1–3.

Bosworth, B. (2016, June 16). South Africa has been key to putting informal settlements on the Habitat III agenda. *Citiscope*. Retrieved from http://citiscope.org/habitatIII/news/2016/06/south-africa-has-been-key-putting-informal-settlements-habitat-iii-agenda.

Bouteligier, S. (2012). *Global cities and networks for global environmental governance*. Hoboken: Taylor and Francis.

Bouteligier, S. (2013). Inequality in new global governance arrangements: The north-south divide in transnational municipal networks. *Innovation—The European Journal of Social Science Research, 26*(3), 251–267. https://doi.org/10.1080/13511610.2013.771890.

Brauch, H., Dalby, S., & Spring, Ú. (2011). Political geoecology for the anthropocene. In H. G. Brauch, Ú. Oswald Spring, C. Mesjasz, J. Grin, P. Kameri-Mbote, B. Chourou, ... J. Birkmann (Eds.), *Coping with global environmental change, disasters and security* (Vol. 5, pp. 1453–1485). Berlin, Heidelberg: Springer.

Braudel, F. (1979). *The wheels of commerce: Civilization and capitalism 15th–18th century*. New York: Harper & Row.

Brenner, N. (1997). State territorial restructuring and the production of spatial scale: Urban and regional planning in the Federal Republic of Germany, 1960–1990. *Political Geography, 16*(4), 273–306.

Brenner, N. (2004). Urban governance and the production of new state spaces in Western Europe, 1960–2000. *Review of International Political Economy, 11*(3), 447–488. https://doi.org/10.2307/4177507.

Brenner, N. (2014). *Implosions/explosions: Towards a study of planetary urbanization*. Berlin: Jovis.

Brenner, N., & Schmid, C. (2011). Planetary urbanisation. In M. Gandy (Ed.), *Urban constellations*. Berlin: Jovis.

Brennetot, A. (2015). The geographical and ethical origins of neoliberalism: The Walter Lippmann Colloquium and the foundations of a new geopolitical order. *Political Geography, 49*, 30–39. https://doi.org/10.1016/j.polgeo.2015.06.007.

Broadhead, L.-A. (2002). *International environmental politics: The limits to green diplomacy*. Boulder, CO: Lynne Rienner.

Brooks, C. (2015). The apostle of internationalism: Stephen Duggan and the geopolitics of international education. *Political Geography, 49*, 64–73. https://doi.org/10.1016/j.polgeo.2015.05.007.

Brown, E. (2017, July 6). *Message to the global citizen festival in Hamburg, Germany*. Retrieved from http://www.latimes.com/politics/essential/la-pol-ca-jerry-brown-hamburg-climate-summit-htmlstory.html.

Brundtland, G. (1991, February 14). Environmental challenges of the 1990s: Our responsibility toward future generations. *The tanner lectures on human values*. Retrieved from https://tannerlectures.utah.edu/_documents/a-to-z/b/brundtland92.pdf.

Bulkeley, H. (2005). Reconfiguring environmental governance: Towards a politics of scales and networks. *Political Geography, 24*(8), 875–902. http://dx.doi.org/10.1016/j.polgeo.2005.07.002.

Bullard, N. (2011). It's too late for sustainability. What we need is system change. *Development, 54*(2), 141–142.

Burke, P., Nelson, G., & Rickerson, W. (2007). Boston's green affordable housing program: Challenges and opportunities. *New York University Journal of Legislation and Public Policy, 11*(1), 1–32.

Burke, R. (2017). The internationalism of human rights. In G. Sluga & P. Clavin (Eds.), *Internationalisms: A twentieth-century history*. Cambridge, UK: Cambridge University Press.

Busch, H. (2015). Linked for action? An analysis of transnational municipal climate networks in Germany. *International Journal of Urban Sustainable Development*, 1–19. https://doi.org/10.1080/19463138.2015.1057144.

Busch, H. (2018). *Entangled cities: Transnational municipal climate networks and urban governance*. Ph.D. Lund University, Lund, Sweden.

Calder, K., & Freytas, M. (2009). Global political cities as actors in twenty-first century international affairs. *SAIS Review of International Affairs, 29*(2), 79.

Cape Town City Council. (1998). *The transportation plan for the Wetton-Landsdowne Corridor. Executive summary*. Cape Town: Unpublished Technical Report, City Planner's Department.

Caprotti, F., Cowley, R., Datta, A., Broto, V. C., Gao, E., Georgeson, L., Herrick, C., Odendaal, N., Joss, S. (2017). The new urban agenda: Key opportunities and challenges for policy and practice. *Urban Research and Practice, 10*(3), 367–378.

Caradonna, J. L. (2014). *Sustainability: A history*. New York: Oxford University Press.

Castán Broto, V., & Bulkeley, H. (2013). A survey of urban climate change experiments in 100 cities. *Global Environmental Change, 23*(1), 92–102. http://dx.doi.org/10.1016/j.gloenvcha.2012.07.005.

Chan, D. K. H. (2016). City diplomacy and "glocal" governance: Revitalizing cosmopolitan democracy. *Innovation—The European Journal of Social Science Research, 29*(2), 134–160. https://doi.org/10.1080/13511610.2016.1157684.

Chatterton, P., Featherstone, D., & Routledge, P. (2013). Articulating climate justice in Copenhagen: Antagonism, the commons, and solidarity. *Antipode, 45*(3), 602–620. https://doi.org/10.1111/j.1467-8330.2012.01025.x.

Chesney, C. (2008). *African-American environmentalism: Issues and trends for teaching, research and extension.* Paper presented at the The Black Environmental Thought: Land, Power and Sustainability Conference, Kellogg Conference Center, Tuskegee University, Tuskegee, Alabama. Retrieved from http://e-research.tnstate.edu/extensionw/1.

Church, J. (2017). Civil society and the open data movement. *DTTP, Documents to the People, 45*(3), 5–8.

City of Cape Town. (2012). *Cape Town spatial development framework: Statutory report.* Cape Town. Retrieved from http://www.capetown.gov.za/work%20and%20business/planning-portal/regulations-and-legislations/cape-town-spatial-development-framework.

City of Cape Town. (2016). *The City of Cape Town's Digital Journey.* Cape Town. Retrieved from http://acceleratecapetown.co.za/wp/wp-content/uploads/2016/11/The-City-of-Cape-Towns-Digital-Journey-Towards-a-Smarter-Future-Rudy-Abrahams-CoCT.pdf..

City of Los Angeles. (2016). *Sustainable city plan: 2nd annual report 2016–2017.* City of Los Angeles. Retrieved from http://plan.lamayor.org/.

City of Seattle. (2016). *Open data program: 2016 annual report.* Seattle: City of Seattle.

Clapp, J., & Dauvergne, P. (2008). *Paths to a green world: The political economy of the global environment.* Cambridge, MA: MIT Press.

Clark, T. P. (2015). Sustainability for whom? An examination of the equity pillar of sustainability at the local policymaking level. *Environmental Justice, 8*(2), 57–63. https://doi.org/10.1089/env.2015.0002.

Clavel, P. (1986). *The progressive city.* New Brunswick: Rutgers University Press.

Clavin, P. (2017). Men and markets: Global capital and the international economy. In G. Sluga & P. Clavel (Eds.), *Internationalisms: A twentieth-century history.* Cambridge, UK: Cambridge University Press.

Corbiere-Nicollier, T., Ferrari, Y., Jemelin, C., & Jolliet, O. (2003). Assessing sustainability: An assessment framework to evaluate Agenda 21 actions at the local level. *International Journal of Sustainable Development and World Ecology, 10*(3), 225–237.

Curtis, S. (Ed.). (2014). *The power of cities in international relations.* New York: Routledge.

Curtis, S. (2016). Cities and global governance: State failure or a new global order? *Millennium, 44*(3), 455–477. https://doi.org/10.1177/0305829816637233.

Czarniawska-Joerges, B. (2002). *A tale of three cities: Or the globalization of city management*. Oxford: Oxford University Press.

Datta, A. (2015). New urban utopias of postcolonial India: 'Entrepreneurial urbanization' in Dholera smart city, Gujarat. *Dialogues in Human Geography, 5*(1), 3–22. https://doi.org/10.1177/2043820614565748.

Davenport, C., & Nagourney, A. (2017, May 23). Fighting Trump on climate, California becomes a global force. *New York Times*. Retrieved from https://www.nytimes.com/2017/05/23/us/california-engages-world-and-fights-washington-on-climate-change.html.

Davidson, K., & Gleeson, B. (2015). Interrogating urban climate leadership: Toward a political ecology of the C40 network. *Global Environmental Politics, 15*(4), 21–38. https://doi.org/10.1162/glep_a_00321.

Davis, M. (1999). *Ecology of fear: Los Angeles and the imagination of disaster*. New York: Vintage Books.

Davis, M. (2006). *City of quartz : Excavating the future in Los Angeles*. London and New York: Verso.

Davis, R. (2013). *Is the future of Cape Town 25km from Cape Town?*. Retrieved from https://www.dailymaverick.co.za/article/2013-05-17-is-the-future-of-capetown-25km-from-cape-town/.

de Jong, M., Joss, S., Schraven, D., Zhan, C. J., & Weijnen, M. (2015). Sustainable-smart-resilient-low carbon-eco-knowledge cities; making sense of a multitude of concepts promoting sustainable urbanization. *Journal of Cleaner Production, 109*, 25–38. https://doi.org/10.1016/j.jclepro.2015.02.004.

de Macedo, L. V., Setzer, J., & Rei, F. (2016). Transnational action fostering climate protection in the city of Sao Paulo and beyond. *Disp, 52*(2), 35–44.

Del Biaggio, C. (2011). Bridging national boundaries: How networks of local actors are building the Alpine region. *Procedia—Social and Behavioral Sciences, 14*, 121–128. https://doi.org/10.1016/j.sbspro.2011.03.029.

Delaney, B. (2017, August 16). Melbourne is 'most liveable city' again. But it's also harder, crueller, out of reach. *The Guardian*. Retrieved from https://amp.theguardian.com/commentisfree/2017/aug/17/melbourne-is-most-liveable-city-again-but-its-also-harder-crueller-out-of-reach.

Dewar, D. (1979). *Low income housing policy in South Africa: With particular reference to the Western Cape/David Dewar, George Ellis*. Cape Town: Urban Problems Research Unit, University of Cape Town.

Dewar, D. (1998, February 8). Personal communication.

Dewar, D., & Todeschini, F. (1998). *Urban integration and economic development*. London: Frankolin.

Dewar, D., & Watson, V. (1990). *The structure and form of metropolitan Cape Town: Its origins, influences and performance* (Working Paper No. 42). Retrieved from Cape Town.

Didier, S., Morange, M., & Peyroux, E. (2013). The adaptative nature of neoliberalism at the local scale: Fifteen years of city improvement districts in Cape Town and Johannesburg. *Antipode, 45*(1), 121–139. https://doi.org/10.1111/j.1467-8330.2012.00987.x.

Dierwechter, Y. (2008). *Urban growth management and its discontents: Promises, practices and geopolitics in US city-regions*. New York: Palgrave.

Dierwechter, Y. (2010). Metropolitan geographies of US climate action: Cities, suburbs, and the local divide in global responsibilities. *Journal of Environmental Policy & Planning, 12*(1), 59–82.

Dierwechter, Y. (2017a). The smart state as utopian space for urban politics. In A. Jonas, B. Miller, K. Ward, & D. Wilson (Eds.), *The Routledge handbook on spaces of urban politics*. London and New York: Routledge.

Dierwechter, Y. (2017b). *Urban sustainability through smart growth: Greater Seattle, intercurrence, and the new geography of regional development*. New York: Springer.

Dierwechter, Y. (2017c). *Urban sustainability through smart growth: Intercurrence, planning, and the geographies of regional development across Greater Seattle*. Cham, Switzerland: Springer.

Dierwechter, Y. (2018). The smart state as utopian space for urban politics. In K. Ward, A. Jonas, B. Miller, & D. Wilson (Eds.), *The Routledge handbook on spaces of urban politics*. New York: Routledge.

Dierwechter, Y., & Wessells, A. (2013). The uneven localisation of climate action in metropolitan Seattle. *Urban Studies, 50*(7), 1368–1385.

Dilworth, R. (Ed.). (2009). *The city in American political development*. New York: Routledge.

Dorogovtsev, S. N., Goltsev, A. V., & Mendes, J. F. F. (2006). K-Core organization of complex networks. *Physical Review Letters, 96*(4), 040601.

Dougherty, C., & Plumber, B. (2018, March 18). A bold, divisive plan to wean Californians from theirs cars. *New York Times*. Retrieved from https://www.nytimes.com/2018/03/16/business/energy-environment/climate-density.html.

Dreier, P., Mollenkopf, J. H., & Swanstrom, T. (2014). *Place matters: Metropolitics for the twenty-first century* (3rd ed.). Lawrence, KS: University Press of Kansas.

Duffield, J. S., & Woodall, B. (2011). Japan's new basic energy plan. *Energy Policy, 39*(6), 3741–3749. https://doi.org/10.1016/j.enpol.2011.04.002.

Duit, A. (Ed.). (2014). *State and environment: The comparative study of environmental governance*. Boston: MIT press.

Eardley-Pryor, R. (2014). *The global environmental moment: Sovereignty and American science on spaceship earth, 1945–1974*. Ph.D. dissertation, University of California, Santa Barbara. Retrieved from https://alexandria.ucsb.edu/downloads/n870zq96x.

Eckersley, R. (Ed.). (2004). *The green state*. Boston: MIT Press.

Engelke, P. (2013). *Foreign policy for an urban world: Global governance and the rise of cities*. Washington, DC: Atlantic Council.

Evans, P. (Ed.) (2002). *Livable cities? Urban struggles for livelihood and Sustainability*. Berkeley University of California Press.

Evans, J., & Karvonen, A. (2014). 'Give me a laboratory and I will lower your carbon footprint!'—Urban laboratories and the governance of low-carbon futures. *International Journal of Urban and Regional Research, 38*(2), 413–430. https://doi.org/10.1111/1468-2427.12077.

Featherstone, D. (2012). *Solidarity: Hidden histories and geographies of internationalism*. London and New York: Zed Books.

Featherstone, D. (2015). Maritime labour and subaltern geographies of internationalism: Black internationalist seafarers' organising in the inter-war period. *Political Geography, 49*, 7–16. https://doi.org/10.1016/j.polgeo.2015.08.004.

Feinberg, T. (2003). *California's workforce development system*. Sacramento, CA: State of California.

Ferguson, B. C., Brown, R. R., Frantzeskaki, N., de Haan, F. J., & Deletic, A. (2013). The enabling institutional context for integrated water management: Lessons from Melbourne. *Water Research, 47*(20), 7300–7314. https://doi.org/10.1016/j.watres.2013.09.045.

Ferretti, F. (2018). Geographies of internationalism: Radical development and critical geopolitics from the Northeast of Brazil. *Political Geography, 63*, 10–19. https://doi.org/10.1016/j.polgeo.2017.11.004.

Forman, M. (1998). *Nationalism and the international labor movement: The idea of the nation in socialist and anarchist theory*. University Park, PA: Pennsylvania State University Press.

Foster, J. (2008). *The sustainability mirage*. London: Earthscan.

Foster, J., Clark, B., & York, R. (2010). *The ecological rift: Capitalism's war on the earth*. New York: Monthly Review Press.

Frankowski, P. (2010). One world-many orders? In S. Krishna-Hensel (Ed.), *Order and disorder in the international system*. Farnham: Taylor and Francis.

Freund, B. (2001). Brown and green in Durban: The evolution of environmental policy in a post-apartheid city. *International Journal of Urban and Regional Research, 25*(4), 717–739.

Frieden, B. (1979). The new regulation comes to suburbia. *The Public Interest, 55*, 15–27.

Frug, G. E. (2008). *City bound: How states stifle urban innovation.* Ithaca: Cornell University Press.
Fukuyama, F. (1992). *The end of history and the last man.* New York: Maxwell Macmillan International.
Fukuyama, F. (2004). *State-building: Governance and world order in the 21st century.* Ithaca, NY: Cornell University Press.
Fukuyama, F. (2013). *Political order and political decay.* New York: Farrar, Strauss & Giroux.
Funtowicz, S. O., Ravetz, J. R., & O'Connor, M. (1998). Challenges in the utilisation of science for sustainable development. *International Journal of Sustainable Development, 1*(1), 2–10.
Garcia-Sanchez, I. M., & Prado-Lorenzo, J. M. (2008). Determinant factors in the degree of implementation of Local Agenda 21 in the European Union. *Sustainable Development, 16*(1), 17–34. https://doi.org/10.1002/sd.334.
Gebre-Egziabher, A. (2004). Sustainable cities programme: A joint UN-HABITAT-UNEP facility on the urban environment with participation of the Dutch government. *Annals of the New York Academy of Sciences, 1023*(1), 62–79. https://doi.org/10.1196/annals.1319.016.
George, S. (1999). *A short history of neoliberalism: Twenty years of elite economics and emerging opportunities for structural change.* Paper presented at the Conference on Economic Sovereignty in a Globalizing World, Bangkok.
Gilpin, R. (1987). *The political economy of international relations.* Princeton, NJ: Princeton University Press.
Girardet, H. (1999). *Creating sustainable cities.* Foxhole, UK: Green Books Ltd.
Glacken, C. (1973). *Traces on the Rhodian shore: Nature and culture in Western thought from ancient times to the end of the eighteenth century.* Berkeley: University of California Press.
Glassman, J. (2009). Critical geography I: The question of internationalism. *Progress in Human Geography, 33*(5), 685–692. https://doi.org/10.1177/0309132508101602.
Goodchild, M. F. (2007). Citizens as sensors: The world of volunteered geography. *GeoJournal, 69*(4), 211–221. https://doi.org/10.1007/s10708-007-9111-y.
Gopalakrishnan, C. (2014). Poem: On sustainability. *Journal of Natural Resources Policy Research, 6*(2–3), 213–213. https://doi.org/10.1080/19390459.2014.924250.
Greenberg, S. (2009). *Sustainable streets: Foundations for an emerging practice.* Paper presented at the Transportation Research Board 88th Annual Meeting, Washington, DC.
Gregory, D. (2000). Dialectic(s). In R. Johnston, D. Gregory, G. Pratt, & M. Watts (Eds.), *The dictionary of human geography* (4th ed., pp. 172–173). Oxford: Blackwell.

Gunder, M. (2010). Planning as the ideology of (neoliberal) space. *Planning Theory, 9*(4), 298–314.

Hackworth, J. (2007). *The neoliberal city: Governance, ideology, and development in American urbanism*. Ithaca: Cornell University Press.

Halperin, M. (1974). *Bureaucratic politics and foreign policy*. Washington, DC: Brookings.

Harari, Y. N. (2015). *Sapiens: A brief history of humankind* (First U.S. edition. ed.). New York: Harper.

Hardt, H., & Negri, A. (2000). *Empire*. Cambridge, MA: Harvard University Press.

Harlow, J., Golub, A., & Allenby, B. (2013). A review of utopian themes in sustainable development discourse. *Sustainable Development, 21*(4), 270–280. https://doi.org/10.1002/sd.522.

Harvey, D. (1996). *Justice, nature and the geography of difference*. Oxford: Blackwell.

Hayman, P., & Williams, J. (2006). Westphalian sovereignty: Rights, intervention, meaning and context. *Global Society, 20*(4), 521–541. https://doi.org/10.1080/13600820600929879.

Healey, P. (2006). *Collaborative planning: Shaping places in fragmented societies*. Basingstoke, Hampshire; New York: Palgrave Macmillan.

Herod, A. (2004). Geographies of labor internationalism. *Social Science History, 27*(4), 501–523. https://doi.org/10.1017/s0145553200012669.

Herrschel, T. (2013). *Cities, state and globalisation. City-regional governance in Europe and North America*. London: Routledge.

Herrschel, T., & Dierwechter, Y. (2015). Smart city-regional governance: A dual transition. *Regions, 300*(4), 20–22.

Herrschel, T., & Dierwechter, Y. (2018). *Smart transitions in city-regionalism: The quest for competitiveness and sustainability*. London: Routledge.

Herrschel, T., & Newman, P. (2017). *Cities as international actors: Urban and regional governance beyond the nation state*. London and New York: Palgrave.

Heyd, T. (1995). Agenda 21 provides a shaky foundation for fight against global poverty. *Alternatives: Perspectives on Society Technology and Environment, 21*(4), 33–33.

Hill, R. C. (2004). Cities and nested hierarchies. *International Social Science Journal, 56*(181), 373–384. https://doi.org/10.1111/j.0020-8701.2004.00500.x.

Hodas, D. (2011). International law and sustainable energy: A portrait in failure. In J. Benidickson, B. Boer, A. Benjamin, & K. Morrow (Eds.), *Environmental law and sustainability after Rio*. Cheltenhem: Edward Elgar.

Hodder, J., Legg, S., & Heffernan, M. (2015). Introduction: Historical geographies of internationalism, 1900–1950. *Political Geography, 49*, 1–6. https://doi.org/10.1016/j.polgeo.2015.09.005.

Hodos, J. (2009). Against expectionalism: Intercurrence and intergovernmental relations in Britain and the United States. In R. Dilworth (Ed.), *The city in American political development* (pp. 44–63). New York: Routledge.

Holden, M., & Scerri, A. (2013). More than this: Liveable Melbourne meets liveable Vancouver. *Cities, 31,* 444–453. https://doi.org/10.1016/j.cities.2012.07.013.

Hollands, R. G. (2015). Critical interventions into the corporate smart city. *Cambridge Journal of Regions, Economy and Society, 8*(1), 61–77. https://doi.org/10.1093/cjres/rsu011.

Imlay, T. (2017). Socialist internationalism after 1914. In G. Sluga & P. Clavin (Eds.), *Internationalisms: A twentieth-century history* (pp. 213–242). Cambridge, UK: Cambridge University Press.

Iriye, A. (2017). Forward. In G. Sluga & P. Clavin (Eds.), *Internationalisms: A twentieth-century history.* Cambridge, UK: Cambridge University Press.

Ivanova, M. (2007). Designing the United Nations Environment Programme: A story of compromise and confrontation. *International Environmental Agreements: Politics, Law and Economics, 7*(4), 337–361. https://doi.org/10.1007/s10784-007-9052-4.

Ivanova, M. (2010). UNEP in global environmental governance: Design, leadership, location. *Global Environmental Politics, 10*(1), 30–59. https://doi.org/10.1162/glep.2010.10.1.30.

Jenkins, P., & Wilkinson, P. (2002). Assessing the growing impact of the global economy on urban development in southern African cities—Case studies in Maputo and Cape Town. *Cities, 19*(1), 33–47. https://doi.org/10.1016/s0264-2751(01)00044-0.

Jin, J., Gubbi, J., Marusic, S., & Palaniswami, M. (2014). An information framework for creating a smart city through internet of things. *IEEE Internet of Things Journal, 1*(2), 112–121. https://doi.org/10.1109/jiot.2013.2296516.

Johnson, R. W. (2015). *How long will South Africa survive?* London: Hurst.

Jonas, A. (2012). City-regionalism as a contingent 'geopolitics of capitalism'. *Geopolitics, 18*(2), 284–298.

Jonas, A. (2013). City-regionalism as a contingent 'geopolitics of capitalism'. *Geopolitics, 18*(2), 284–298. https://doi.org/10.1080/14650045.2012.723290.

Jonas, A. (2015, October 8–9). *Variegated city regionalisms: The geopolitics of city-region building.* Paper presented at the Smart City-Regional Governance, Cosmopolis, Centre for Urban Policy Research, Vrije Universiteit Brussel, Brussels.

Jonas, A. & Moisio, S. (2016). City regionalism as geopolitical processes: A new framework for analysis. *Progress in Human Geography.* https://doi.org/10.1177/0309132516679897.

Jonas, A., While, A., & Gibbs, D. (2004). State modernisation and local strategic selectivity after Local Agenda 21: Evidence from three northern English localities. *Policy & Politics, 32*(2), 151–168. https://doi.org/10.1332/0305573047735581 25.

Jones, K. (2018). Sierra Club opposes SB 827 [Press release]. Retrieved from http://citizenmarin.org/news/sierra-club-opposes-sb-827/.

Kanninen, T. (Ed.). (2012). *Crisis of global sustainability.* London: Routledge.

Keiner, M., & Kim, A. (2007). Transnational city networks for sustainability. *European Planning Studies, 15*(10), 1369–1395. https://doi.org/10.1080/09654310701550843.

Keohane, R. O., & Nye, J. S. (1973). Power and interdependence. *Survival, 15*(4), 158–165. https://doi.org/10.1080/00396337308441409.

Kern, K., & Bulkeley, H. (2009). Cities, Europeanization and multi-level governance: Governing climate change through transnational municipal networks. *Journal of Common Market Studies, 47*(2), 309–332.

Kern, K., Koll, C., & Schophaus, M. (2007). The diffusion of Local Agenda 21 in Germany: Comparing the German federal states. *Environmental Politics, 16*(4), 604–624. https://doi.org/10.1080/09644010701419139.

Khilnani, S. (1997). *The idea of India.* London: Penguin.

Kneebone, E., & Garr, E. (2010). *The suburbanization of poverty: Trends in metropolitan America, 2000 to 2008.* Washington, DC: Brookings Institution.

Kolbert, E. (2014). *The sixth extinction: An unnatural history.* New York: Henry Holt.

Krause, R. M. (2012). An assessment of the impact that participation in local climate networks has on cities' implementation of climate, energy, and transportation policies. *Review of Policy Research, 29*(5), 585–604. https://doi.org/10.1111/j.1541-1338.2012.00582.x.

Krishna-Hensel, S. (2010). *Order and disorder in the international system.* Farnham: Taylor and Francis.

Krueger, R. (2010). Smart growth and its discontents: An examination of American and European approaches to local and regional sustainable development. *Documents d'Anàlisi Geogràfica, 56*(3), 409–433.

Krugman, P. (2017, June 1). Trump gratuitously rejects the Paris Climate Accord. *New York Times.* Retrieved from https://www.nytimes.com/2017/06/01/opinion/trump-gratuitously-rejects-the-paris-climate-accord.html?_r=0.

Kuznetsov, A. S. (2015). *Theory and practice of paradiplomacy subnational governments in international affairs.* London and New York: Routledge.

Lancashire County Council. (2003a). *Local Agenda 21—End of term report.* Retrieved from http://www3.lancashire.gov.uk/council/meetings/displayFile.asp?FTYPE=A&FILEID=3329.

Lancashire County Council. (2003b). *Local Agenda 21—Second annual progress report.* Retrieved from www3.lancashire.gov.uk/council/meetings/displayFile.asp?FTYPE=A&FILEID=439.

Langeweg, F. (1998). The implementation of Agenda 21 'our common failure'? *Science of the Total Environment, 218*(2–3), 227–238. https://doi.org/10.1016/s0048-9697(98)00210-1.

Le Faye, D. (1995). *Jane Austen's letters.* Oxford: Oxford University Press.

Lee, T. (2013). Global cities and transnational climate change networks. *Global Environmental Politics, 13*(1), 108–127. https://doi.org/10.1162/GLEP_a_00156.

Lee, T., & van de Meene, S. (2012). Who teaches and who learns? Policy learning through the C40 cities climate network. *Policy Sciences, 45*(3), 199–220. https://doi.org/10.1007/s11077-012-9159-5.

Lefebvre, H. (1991). *The production of space.* Oxford, UK: Blackwell.

Lefebvre, H. (2013). *Rhythmanalysis: Space, time, and everyday life.* New York: Bloomsbury.

Leonard, L. (2010). *Global ecological politics.* Bingley: Emerald Group.

Lindblom, C. E. (1977). *Politics and markets: The world's political economic systems.* New York: Basic Books.

Lipsett, S., & Marks, G. (2000). *It didn't happen here: Why socialism failed in the United States.* New York: W. W. Norton.

Llamas-Sanchez, R., Munoz-Fernandez, A., & Maraver-Tarifa, G. (2011). The local Agenda 21 in Andalusia, Spain: A model for sustainable innovation. *African Journal of Business Management, 5*(32), 12653–12663. https://doi.org/10.5897/ajbm11.2381.

Lombardi, P., Giordano, S., Farouh, H., & Yousef, W. (2012). Modelling the smart city performance. *Innovation—The European Journal of Social Science Research, 25*(2), 137–149. https://doi.org/10.1080/13511610.2012.660325.

Lucas, J. (2015). Urban governance and the American political development approach. *Urban Affairs Review.* https://doi.org/10.1177/1078087415620054.

Lumley, S., & Armstrong, P. (2004). Some of the nineteenth century origins of the sustainability concept. *Environment, Development and Sustainability, 6,* 367–378.

Ma, X. L., Liu, C. C., Wen, H. M., Wang, Y. P., & Wu, Y. J. (2017). Understanding commuting patterns using transit smart card data. *Journal of Transport Geography, 58,* 135–145. https://doi.org/10.1016/j.jtrangeo.2016.12.001.

MacLeod, G. (2013). New urbanism/smart growth in the Scottish Highlands: Mobile policies and post-politics in local development planning. *Urban Studies, 50*(11), 2196–2221. https://doi.org/10.1177/0042098013491164.

Macleod, G., & Goodwin, M. (1999). Space, scale and state strategy: Rethinking urban and regional governance. *Progress in Human Geography, 23*(4), 503–527.

Magee, L. (2016). *Interwoven cities*. Houndmills, Basingstoke, Hampshire; New York, NY: Palgrave Macmillan.

Maher, N. (2008). *Nature's new deal*. New York: Oxford University Press.

Massey, D. (1993). Power-geometries and a progressive sense of place. In J. Bird (Ed.), *Mapping the futures: Local cultures, global change*. London and New York: Routledge.

Mathews, J. T. (1997). Power shift. (changing role of central government). *Foreign Affairs, 76*(1), 50. https://doi.org/10.2307/20047909.

Mayer, J. (1997). Gifford Pinchot, John Muir, and the boundaries of politics in American thought. *Polity, 2*(2), 267–284.

Mazower, M. (2012). *Governing the world: The history of an idea, 1815 to the present*. New York: The Penguin Press.

McDonald, C., Frost, L., Rainnie, A., & Kirk-Brown, A. (2008). *The new regionalism and the role of intermediary organisations in regional development*. Paper presented at the Regional Studies Association Annual Conference, Prague, Czech Republic.

Meadowcroft, J. (2011). Engaging with the politics of sustainability transitions. *Environmental Innovation and Societal Transitions, 1*, 70–75.

Meadows, D. (2008). *30-Year update of limits to growth finds global society in "overshoot," foresees social, economic, and environmental decline*. Retrieved from https://web.archive.org/web/20080517054744, http://www.clubofrome.org/docs/confs/meadows_abstract_21_08_04.pdf.

Merrifield, A. (2013). The urban question under planetary urbanization. *International Journal of Urban and Regional Research, 37*(3), 909–922. https://doi.org/10.1111/j.1468-2427.2012.01189.x.

Miller, D. L. (1989). *Lewis Mumford, a life*. New York: Weidenfeld & Nicolson.

Mohieldin, M., & Ijjasz-Vasquez, E. (2018). *Localizing the SDGs in Colombia, Indonesia and Kenya*. Retrieved from http://www.citiscope.org/commentary/2017/10/localizing-sdgs-colombia-indonesia-and-kenya.

Moisio, S. (2018). Urbanizing the nation-state? Notes on the geopolitical growth of cities and city-regions. *Urban Geography*, 1–4. https://doi.org/10.1080/02723638.2018.1454685.

Mol, A., & Janicke, M. (2009). The origins and theoretical foundations of ecological modernisation theory. In A. Mol, D. Sonnefeld, & G. Spargaren (Eds.), *The ecological modernisation reader* (pp. 17–27). London: Routledge.

Momtaz, D. (1996). The United Nations and the protection of the environment: From Stockholm to Rio de Janeiro. *Political Geography, 15*(3), 261–271. http://dx.doi.org/10.1016/0962-6298(95)00109-3.

Monbiot, G. (2000). *Captive state: The corporate takeover of Britain*. London: Macmillan.

Moore, S. (2007). *Alternative routes to the sustainable city: Austin, Curitiba, and Frankfurt*. Lanham, MD: Lexington Books.

Morandi, C., Rolando, A., & Di Vita, S. (2016). *From smart city to smart region: Digital services for an internet of places*. Cham: Springer.

Morgan, K. (2007). The polycentric state: New spaces of empowerment and engagement? *Regional Studies, 41*(9), 1237–1251. https://doi.org/10.1080/00343400701543363.

Mossner, S., & Miller, B. (2015). Sustainability in one place? Dilemmas of sustainability governance in the Frieburg Metropolitan region. *Regions, 300*(Winter), 19–21.

Murphy, J., & Gouldson, A. (2000). Environmental policy and industrial innovation: Integrating environment and economy through ecological modernisation. *Geoforum, 31*(1), 33–44. https://doi.org/10.1016/S0016-7185(99)00042-1.

Neirotti, P., De Marco, A., Cagliano, A. C., Mangano, G., & Scorrano, F. (2014). Current trends in smart city initiatives: Some stylised facts. *Cities, 38*, 25–36. https://doi.org/10.1016/j.cities.2013.12.010.

Newman, E. (2009). Failed states and international order: Constructing a post-Westphalian world. *Contemporary Security Policy, 30*(3), 421–443. https://doi.org/10.1080/13523260903326479.

Noortmann, M., Reinisch, A., & Ryngaert, C. (2015). *Non-state actors in international law*. Oxford; Portland, Oregon: Hart Publishing.

Nwaka, G. I. (1996). Planning sustainable cities in Africa. *Canadian Journal of Urban Research, 5*(1), 119–136.

OECD. (2002). *Learning to innovate: Learning regions*. Paris: OECD.

OECD. (2008). *Territorial review: Cape Town metropolitan review*. Paris: OECD.

Ogbazi, J. U. (2013). Alternative planning approaches and the sustainable cities programme in Nigeria. *Habitat International, 40*, 109–118. https://doi.org/10.1016/j.habitatint.2013.03.001.

Orren, K., & Skowronek, S. (1996). Institutions and intercurrence: Theory building in the fullness of time. *Nomos XXXVII, Political Order, 38*, 111–146.

Orren, K., & Skowronek, S. (2004). *The search for American political development*. Cambridge: Cambridge University Press.

Osofsky, H. (2005). The geography of climate change litigation: Implications for transnational regulatory governance. *Washington University Law Quarterly, 83*(6), 1789–1856.

Osofsky, H. (2015). Rethinking the geography of local climate action: Multilevel network participation in metropolitan regions. *Utah Law Review, 15*(1), 173–240.

Parnell, S., & Robinson, J. (2012). (Re)Theorizing cities from the global south: Looking beyond neoliberalism. *Urban Geography, 33*(4), 593–617.

Peck, J. (2008). Remaking laissez-faire. *Progress in Human Geography, 32*(1), 3–43. https://doi.org/10.1177/0309132507084816.

Perlman, B. J., & Jimenez, J. (2010). Creative regionalism: Governance for stressful times. *State and Local Government Review, 42*(2), 151–155. https://doi.org/10.1177/0160323x10380616.

Portney, K. E. (2003). *Taking sustainability seriously: Economic development, the environment, and quality of life in American cities.* Boston: MIT Press.

Pred, A. (1977). The choreography of existence: Comments on Hagerstrand's time-geography and its usefulness. *Economic Geography, 53*, 207–221.

Preston, P. (1996). *Development theory: An introduction.* Oxford: Blackwell.

Pycroft, C. (1998). Integrated development planning or strategic paralysis? Municipal development during the local government transition and beyond. *Development Southern Africa, 15*(2), 151–163.

Rashidi, K., & Patt, A. (2018). Subsistence over symbolism: The role of transnational municipal networks on cities' climate policy innovation and adoption. *Mitigation and Adaptation Strategies for Global Change, 23*(4), 507–523. https://doi.org/10.1007/s11027-017-9747-y.

Real, W. (1999). Challenges of Agenda 21 to the environmental education. *Fresenius Environmental Bulletin, 8*(5–6), 230–235.

Reuters. (2016, June 18). California passes France as world's 6th-largest economy. *Rueters.* Retrieved from http://fortune.com/2016/06/17/california-france-6th-largest-economy/.

Roberts, D., & Diederichs, N. (2002). Durban's Local Agenda 21 programme: Tackling sustainable development in a post-apartheid city. *Environment and Urbanization, 14*(1), 189–201.

Robertson, T. (2012). *The Malthusian moment: Global population growth and the birth of American environmentalism.* New Brunswick: Rutgers University Press.

Robinson, J. (2006). *Ordinary cities: Between modernity and development.* London and New York: Routledge.

Rodríguez-Pose, A. (2018). The revenge of the places that don't matter (and what to do about it). *Cambridge Journal of Regions, Economy and Society, 11*(1), 189–209.

Rodríguez-Pose, A., & Crescenzi, R. (2008). Mountains in a flat world: Why proximity still matters for the location of economic activity. *Cambridge Journal of Regions, Economy and Society, 1*(3), 371–388. https://doi.org/10.1093/cjres/rsn011.

Rogers, R., Jalal, K., & Boyd, J. (2005). *An introduction to sustainable development.* Cambridge, MA: London Harvard University Press.

Rogerson, C. (2004). From spatial development initiative to blue IQ: Subnational economic planning in Gauteng. *Urban Forum, 15*(1), 74–101.

Rogerson, C. (2009). The turn to 'new regionalism': South African reflections. *Urban Forum, 20*(2), 111–140. https://doi.org/10.1007/s12132-009-9057-x.

Rogerson, C., & Rogerson, J. (2015). Johannesburg 2030: The economic contours of a "linking global city". *American Behavioral Scientist, 59*(3), 347–368. https://doi.org/10.1177/0002764214550303.

Ruffin, F. (2013). Municipal international relations: The South African case of metropolitan eThekwini. *Loyola Journal of Social Sciences, XXVII*(1), 119–141.

Sarkar, J. (2015). Debating a post-Westphalian international order. *Mainstream Weekly, LIII*(15).

Sassen, S. (1991). *The global city: New York, London, Tokyo*. Princeton: Princeton University Press.

Savitch, H. V., & Adhikari, S. (2017). Fragmented regionalism. *Journal of Urban Affairs, 53*(2). https://doi.org/10.1177/1078087416630626.

Schroeder, H., & Bulkeley, H. (2009). Global cities and the governance of climate change: What is the role of law in cities. *Fordham Urban Law Journal, 36*, 314–360.

Scott, A. (2001). Globalization and the rise of city regions. *European Planning Studies, 9*(7), 813–826.

Scott, A. (Ed.). (2001). *Global city-regions: Trends, theory, policy*. Oxford: Oxford University Press.

Scott, J. W. (2007). Smart growth as urban reform: A pragmatic 'recoding' of the new regionalism. *Urban Studies, 44*(1), 15–35.

Scruggs, G. (2017). *How Anne Hidalgo plans to get cities to adopt climate action plans by 2020*. Retrieved from http://citiscope.org/story/2017/how-anne-hidalgo-plans-get-cities-adopt-climate-action-plans-2020.

Seabrooke, L. (2002). *Bringing legitimacy back in to neo-Weberian state theory and international relations* (Working Paper No. 6). Department of International Relations. Canberra: Australian National University.

Sellers, J. M. (2002). *Governing from below: Urban regions and the global economy*. Cambridge and New York: Cambridge University Press.

Setzer, J. (2015). Testing the boundaries of subnational diplomacy: The international climate action of local and regional governments. *Transnational Environmental Law, 4*(2), 319–337. https://doi.org/10.1017/s2047102515000126.

Skocpol, T. (1985). Introduction. In P. Evans, R. Rueschemeyer, & T. Skocpol (Eds.), *Bringing the state back in* (pp. 3–38). Cambridge and New York: Cambridge University Press.

Skowronek, S. (1982). *Building a new American state: The expansion of national administrative capacities, 1877–1920*. Cambridge [Cambridgeshire] and New York: Cambridge University Press.

Slaughter, A.-M. (1997). The real new world order. *Foreign Affairs, Sept/Oct*, 183–197.

Sluga, G., & Clavin, P. (2017a). Rethinking the history of internationalism. In G. Sluga & P. Clavin (Eds.), *Internationalisms: A twentieth century history* (pp. 3–16). Cambridge: Cambridge University Press.

Sluga, G., & Clavin, P. (Eds.). (2017b). *Internationalisms: A twentieth-century history*. Cambridge, UK: Cambridge University Press.

Smart Cities Council. (2016, September 15). About us. *Smart Cities Council*. Retrieved from https://smartcitiescouncil.com/article/about-us-global.

Smith, M. (2001). *Transnational urbanism: Locating globalization*. Malden, MA: Blackwell.

Smith, R. (2003). World city actor-networks. *Progress in Human Geography, 27*(1), 25–44.

Soja, E. (2000). *Postmetropolis*. London and New York: Sage.

Soja, E. (2010). *Seeking spatial justice*. Minneapolis: University of Minnesota Press.

Sowell, T. (1998). *Conquests and cultures: An international history* (1st ed.). New York: Basic Books.

Sowman, M., & Brown, A. L. (2006). Mainstreaming environmental sustainability into South Africa's integrated development planning process. *Journal of Environmental Planning & Management, 49*(5), 695–712. https://doi.org/10.1080/09640560600849988.

State Capacity Research Project. (2017). *Betrayal of the promise: How South Africa is being stolen*. Retrieved from https://pari.org.za/wp-content/uploads/2017/05/Betrayal-of-the-Promise-25052017.pdf.

Steffek, J. (2015). Fascist internationalism. *Millennium, 44*(1), 3–22. https://doi.org/10.1177/0305829815581870.

Steger, M. (2002). *Globalism*. Landham, MD: Rowman and Littlefield.

Stone, C., & Whelan, R. (2009). Through the glass darkly: The once and future study of urban politics. In R. Dilworth (Ed.), *The city in American political development*. New York: Routledge.

Storper, M. (2018). Separate worlds? Explaining the current wave of regional economic polarization. *Journal of Economic Geography, 18*(2), 247–270. https://doi.org/10.1093/jeg/lby011.

Suryadevara, N. K., Mukhopadhyay, S. C., Kelly, S. D. T., & Gill, S. P. S. (2015). WSN-based smart sensors and actuator for power management in intelligent buildings. *IEEE/ASME Transactions on Mechatronics, 20*(2), 564–571. https://doi.org/10.1109/tmech.2014.2301716.

Swanstrom, T. (2006). Regionalism, equality, and democracy. *Urban Affairs Review, 42*(2), 249–257.

Swenson, P. (2002). *Capitalists against markets: The making of labor markets and welfare states in the United States and Sweden*. New York: Oxford University Press.

Tavares, R. (2016). *Paradiplomacy: Cities and states as global players*. New York: Oxford University Press.

Taylor, P. J. (2004). *World city network: A global urban analysis*. London and New York: Routledge.

Taylor, P. J. (2005). New political geographies: Global civil society and global governance through world city networks. *Political Geography, 24*(6), 703–730.

Taylor, P. J. (2012). Extraordinary cities: Early 'City-ness' and the origins of agriculture and states. *International Journal of Urban and Regional Research, 36*(3), 415–447.

Taylor, P. J. (2013). *Extraordinary cities: Millenia of moral syndromes, world-systems and city/state relations*. Cheltenham: Edward Elgar.

Thibert, J. (2015). *Governing urban regions through collaboration: A view from North America*. Burlington, VT: Ashgate.

Thrift, N., & Amin, A. (2002). *Cities: Reimagining the urban*. Cambridge: Polity.

Toly, N. (2008). Transnational municipal networks in climate politics: From global governance to global politics. *Globalizations, 5*(3), 341–356.

Toly, N. (2017). The new urban agenda and the limits of cities. *The Hedgehog Review, 19*(2), 36–44.

Tomlinson, R. (2017). An argument for metropolitan government in Australia. *Cities, 63*, 149–153. https://doi.org/10.1016/j.cities.2016.10.013.

Törnqvist, G. (2004). Creativity in time and space. *Geografiska Annaler: Series B, Human Geography, 86*(4), 227–243. https://doi.org/10.1111/j.0435-3684.2004.00165.x.

Townsend, A. M. (2013). *Smart cities: Big data, civic hackers, and the quest for a new utopia* (1st ed.). New York: W. W. Norton.

Trapenberg Frick, K. (2013). The actions of discontent: Tea party and property rights activists pushing back against regional planning. *Journal of the American Planning Association, 79*(3), 190–200. https://doi.org/10.1080/01944363.2013.885312.

Trudeau, D. (2013). New urbanism as sustainable development? *Geography Compass, 7*(6), 435–448.

Tulloch, L., & Neilson, D. (2014). *The Neoliberalisation of Sustainability. Citizenship, 13*(1), 26–38.

UN-Habitat. (2012). *History, mandate & role in the UN system*. Retrieved from https://unhabitat.org/history-mandate-role-in-the-un-system/.

United Nations. (2017). *New urban agenda*. Quito: UN Conference of Housing and Sustainable Urban Development. Retrieved from http://habitat3.org/wp-content/uploads/NUA-English.pdf.

United Nations Conference on Environment and Development. (1993). *Agenda 21: Programme of action for sustainable development*. New York, NY: United Nations Department of Public Information.

Van Hamme, F., & Pion, G. (2012). The relevance of the world-system approach in the era of globalization of economic flows and networks. *Geografiska Annaler: Series B, Human Geography, 94*(1), 65–82.

Van Loon, A. (2018, February 3). Let Cape Town revolutionise the way we think about water. *The Guardian (UK)*. Retrieved from https://www.theguardian.com/commentisfree/2018/feb/04/let-cape-town-revolutionise-the-way-we-think-about-water.

Venter, A. (2007). Integrated development planning as an approach to sustainable development. In G. van Der Waldt (Ed.), *Municipal management: Serving the people* (pp. 93–108). Cape Town: JUTA Press.

Vojnovic, I. (2013). Advancing toward urban sustainability: The pursuit of equity. In I. Vojnovic (Ed.), *Urban sustainability: A Global Context* (pp. 1–34). East Lansing: Michigan State University Press.

Walker, R. (2015). Building a better theory of the urban: A response to 'Towards a new epistemology of the urban?' *City, 19*(2–3), 183–191. https://doi.org/10.1080/13604813.2015.1024073.

Wallerstein, I. M. (1979). *The capitalist world-economy: Essays*. Cambridge, UK: Cambridge University Press.

Wang, V., & McKinley, J. (2018, March 31). In New York budget, a fusillade against de Blasio. *New York Times*. Retrieved from https://www.nytimes.com/2018/03/31/nyregion/new-york-city-budget-cuomo.html.

Ward, K., Jonas, A., Miller, B., & Wilson, D. (2018). The spaces of urban politics: An introduction. In K. Ward, A. Jonas, B. Miller, & D. Wilson (Eds.), *The Routledge handbook on spaces of urban politics*. New York: Routledge.

Waterman, P., & Wills, J. (2001). *Place, space and the new labour internationalisms*. Oxford, UK; Malden, MA: Blackwell.

Waterman, P., & Wills, J. (2012). Place, space and the new labor internationalisms: Beyond the fragments? In P. Waterman & J. Wills (Eds.), *Place, space and the new labour internationalisms*. London: Blackwell.

Watson, V. (1991). Urbanization policy: Lessons from South America for South Africa? *Social Dynamics, 17*(2), 155–167.

Watson, V. (2002). *Change and continuity in spatial planning: Metropolitan planning in Cape Town under political transition*. London and New York: Routledge.

Watson, V. (2013). African urban fantasies: Dreams or nightmares? *Environment and Urbanization, 26*(1), 215–231. https://doi.org/10.1177/0956247813513705.

Weiss, T. G. (2012). *What's wrong with the United Nations and how to fix it* (2nd ed.). Cambridge, UK; Malden, MA: Polity.

While, A., Jonas, A., & Gibbs, D. (2004). The environment and the entrepreneurial city: Searching for the urban 'sustainability; fix' in Manchester and Leeds. *International Journal of Urban and Regional Research, 28*(3), 549–569.

Wiig, A. (2015). The empty rhetoric of the smart city: From digital inclusion to economic promotion in Philadelphia. *Urban Geography*, 1–19. https://doi.org/10.1080/02723638.2015.1065686.

Yacobi, H. (2009). Towards urban geopolitics. *Geopolitics, 14*(3), 576–581. https://doi.org/10.1080/14650040802694091.

Yeh, A. (2017). *Smart cities in Pacific Asia: Roles of government and private sectors.* Paper presented at the The Social and Environmental Implications of Smart Cities: Toward a Global Comparative Research Agenda, University of Calgary.

Yigitcanlar, T., O'Connor, K., & Westerman, C. (2008). The making of knowledge cities: Melbourne's knowledge-based urban development experience. *Cities, 25*(2), 63–72. https://doi.org/10.1016/j.cities.2008.01.001.

Zahran, S., Brody, S. D., Vedlitz, A., Grover, H., & Miller, C. (2008). Vulnerability and capacity: Explaining local commitment to climate-change policy. *Environment and Planning C-Government and Policy, 26*(3), 544–562. https://doi.org/10.1068/c2g.

Index

Y. Dierwechter, *The Urbanization of Green Internationalism*, Cities and the Global Politics of the Environment, https://doi.org/10.1007/978-3-030-01015-7

I

J

K

L

M

The manufacturer's authorised representative in the EU is Springer Nature Customer Service Centre GmbH, Europaplatz 3, 69115 Heidelberg, Germany. If you have any concerns regarding our products, please contact ProductSafety@springernature.com

Printed and bound by CPI Group (UK) Ltd, Croydon, CR0 4YY
30/09/2026
02225666-0019